# SpringerBriefs in Earth System Sciences

## South America and the Southern Hemisphere

*Series editors*

Gerrit Lohmann, Bremen, Germany
Lawrence A. Mysak, Montreal, Canada
Justus Notholt, Bremen, Germany
Jorge Rabassa, Ushuaia, Argentina
Vikram Unnithan, Bremen, Germany

For further volumes:
http://www.springer.com/series/10032

Sandra Gordillo · María Sol Bayer
Gabriella Boretto · Melisa Charó

# Mollusk Shells as bio-geo-archives

## Evaluating Environmental Changes During the Quaternary

 Springer

Sandra Gordillo
María Sol Bayer
Gabriella Boretto
Centro de Investigaciones en Ciencias
de la Tierra
CONICET and Universidad Nacional
de Córdoba
Córdoba
Argentina

Melisa Charó
Facultad de Ciencias Naturales
Universidad Nacional de La Plata
CONICET
La Plata
Argentina

ISSN 2191-589X        ISSN 2191-5903   (electronic)
ISBN 978-3-319-03475-1    ISBN 978-3-319-03476-8   (eBook)
DOI 10.1007/978-3-319-03476-8
Springer Cham Heidelberg New York Dordrecht London

Library of Congress Control Number: 2013957875

Printed on acid-free paper

Springer is part of Springer Science+Business Media (www.springer.com)

*This book is dedicated to Geerat J. Vermeij for teaching us through his writing how to read nature's message in shells*

# Preface

This book is the result of 25 years of work focusing on mollusk shells from southern South America. My interest in seashells started when I was a young girl. Although I was born in a city in the center of Argentina, I spent my holidays by the sea, where I discovered the beauty of shells just waiting to be unearthed from the beaches. Amazed by their colors and shapes, I asked myself where they came from, how they got there, and how deep in the sea they lived. After graduating, I decided to find the answers. I moved to Ushuaia, where I started to study Quaternary mollusks with a CONICET fellowship under the supervision of Jorge O. Rabassa and M. Teresa Sánchez[†]. During this adventure many answers were found and new questions arose, and in recent years three doctoral students, María Sol Bayer, Gabriella M. Boretto, and Melisa P. Charó, joined the team to work under my supervision.

As a result of this rich, shared experience, the purpose of this book is to provide an approach to how mollusk shell remains are used in the reconstruction of Quaternary marine environments in southern South America.

This book is designed for researchers who focus on paleoenvironmental reconstruction, and also for undergraduate and postgraduate students who are interested in Quaternary research, and for everyone who has an interest in this topic and/or in this region.

Sandra Gordillo

# Acknowledgments

We are grateful to Jorge O. Rabassa (CADIC, CONICET) for inviting us onto this book project and to Thomas Brey (Alfred Wegener Institute for Polar and Marine Research) and Gianni Zanchetta (University of Pisa) for their research facilities which helped with the isotopic and sclerochronological analyses. We would also like to thank CONICET and FONCYT from Argentina and the DAAD German Academic Exchange Service for their support of our research.

# Contents

# Abstract

Given the wide range of disciplines involved in Quaternary research, this book aims to provide a comprehensive approach to how mollusk shell remains have been used in the reconstruction of Quaternary environments in southern South America. Our study was based on present-day Holocene and Pleistocene mollusk assemblages from different areas covering a wide distribution range between 40° and 54° S. The mollusk assemblages and/or selected taxa were analyzed using a multidisciplinary approach mainly involving taphonomy, paleoecology, morphometry, shell microstructure, and sclerochronology.

A local-scale quantitative and qualitative analysis of mollusk assemblages at different latitudes suggests that each environment acted as a "dynamic mosaic" for the development of local communities in patchy habitats or sub-environments which, in accordance with sea-level changes, shifted over time. Changes in mollusk assemblages took place from a few hundred to several thousand years ago, and mostly follow local physical variations (i.e., substrate, availability of food, and currents); changes associated to global-scale climatic variability during the Holocene were also recorded in individual taxa. As the multi-proxy evidence used in this study provides a consistent picture of spatial and temporal environmental and climatic changes in southern South America, we have concluded that mollusk shells are extremely valuable tools for studies addressing Quaternary environments anywhere.

**Keywords** Southern South America · Quaternary · Pleistocene · Holocene · Mollusca · Taphonomy · Paleoecology · Morphometry · Shell microstructure · Sclerochronology

# Chapter 1
# Introduction

**Abstract** The Quaternary system/period is the most recent geological time interval in the history of the Earth, and covers the last 2.588 million years up to the present day. It includes a series of very extensive environmental changes which have affected and shaped landscapes and life on Earth. These variations in turn have driven rapid changes in both continental and marine biota. With this in mind, the aim of this book is to provide tools for reconstructing Quaternary environments, based on specific cases which offer an approach to the use of mollusk shells as a multi-proxy data source.

**Keywords** Southern South America · Quaternary · Pleistocene · Holocene · Mollusca · Glacial stages · Interglacial stages · Multi-proxy studies

## 1.1 The Quaternary and the Use of Proxy Records

The Quaternary system/period is the most recent geological time interval in the history of the Earth, and covers the last 2.588 million years up to the present day (Fig. 1.1). It includes a series of very extensive environmental changes which have affected and shaped landscapes and life on Earth. In the geologic history of Earth, the Quaternary is a unit of time within the Cenozoic era and includes two formally designed intervals of series/epoch status (Gibbard et al. 2009): the Pleistocene and the Holocene. The Pleistocene began when rock strata show extensive evidence of widespread expansion of ice sheets over the continents, and was the beginning of an era of dramatic climatic and oceanographic changes. The Holocene is generally regarded as having begun 10,000 radiocarbon years before 1950 AD, or 11.7 k calendar years before 2000 AD (Wolff 2008). One of the most distinctive features of the Quaternary has been periodic cold or glacial stages alternating with interglacial stages (Fig. 1.2), during which temperatures were occasionally higher than those of the present day (Lowe and Walker 1997). During and between these glacial periods, rapid changes in climate and sea level occurred, and environments

S. Gordillo et al., *Mollusk Shells as bio-geo-archives*,
South America and the Southern Hemisphere, DOI: 10.1007/978-3-319-03476-8_1,
© The Author(s) 2014

**Fig. 1.1** The Quaternary in relation to the geological timescale according to the 2013 chronostratigraphic chart recommended by the International Commission on Stratigraphy (Cohen et al. 2013)

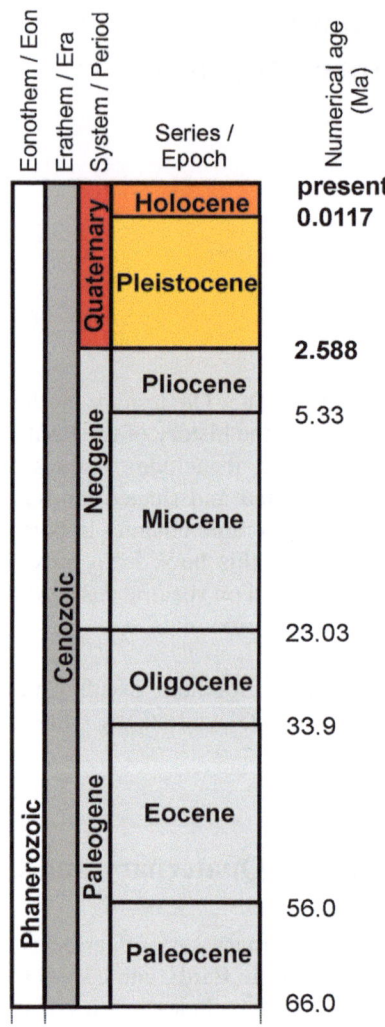

worldwide were altered. These variations in turn have driven rapid changes in both continental and marine biota.

With regard to continental and marine environments, the repeated climatic changes that occurred during the Quaternary have given rise to a rich and complex record of landforms, sediments and biologic remains. These bio-geo-records provide indirect measurements of former climates and environments. When the term bio-geo-archive is associated with mollusk shells, it means that it is possible to reconstruct past ecosystems and environmental conditions from different data sources obtained from shells.

**Fig. 1.2** Curve representing
glacial and interglacial cycles
for the last 2.6 Ma, measured
from oxygen isotope analysis
performed on benthic
foraminifera. Odd numbers
indicate isotope warm stages,
while even numbers indicate
cold stages for the last 0.4 Ma
in Argentina (modified after
Lowe and Walker 1997)

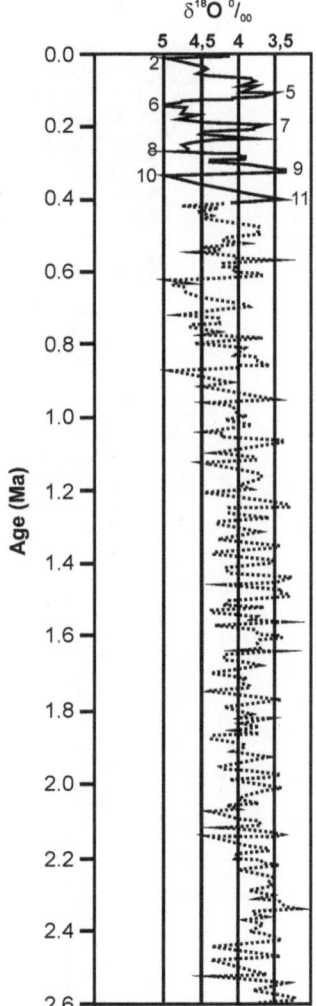

## 1.2  The Scope of this Book

The aim of this book is to provide tools for reconstructing Quaternary environments in southern South America (Fig. 1.3), based on specific cases which offer an approach to the use of mollusk shells as a multi-proxy data source.

On the basis of several examples that include bivalves and chitons, Chap. 2 explains how different taphonomic attributes (fragmentation, dissolution and abrasion, among others) are suitable for evaluating the post-mortem events that took place up to when the mollusk shells were found within the fossil record. In Chap. 3, the shell microstructure of Patagonian aragonitic shells is studied by

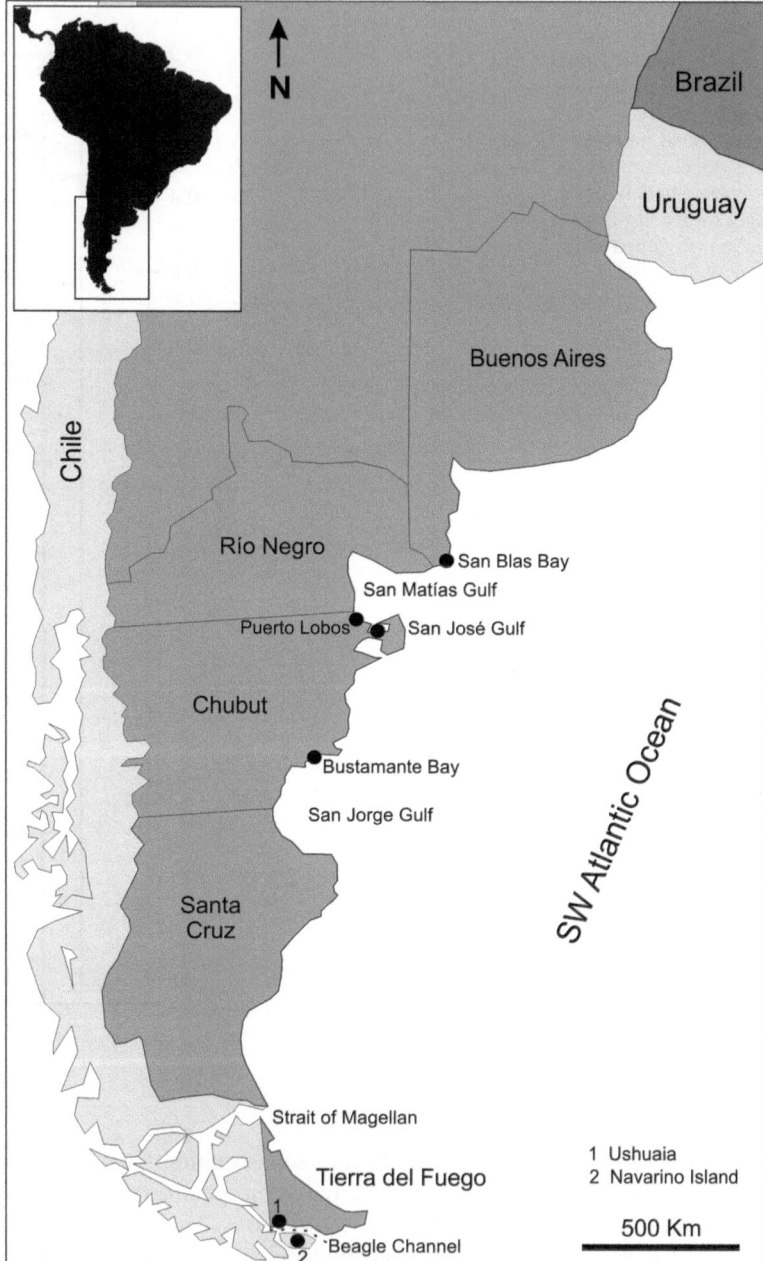

**Fig. 1.3** Location map of southern South America showing the main sites mentioned in the text, covering a wide range between 40° and 54° S

X-ray diffraction, optical and electron microscopy, electron microprobe analyses and microindentation in order to characterize early diagenetic changes and mechanical resistance over a period that exceeds 100,000 years (from the present day up to the late Pleistocene). Chapter 4 is centered on the analysis of taxonomic groups and taxa composition of the preserved fauna from different Pleistocene and Holocene marine outcrops located mainly along the Argentinean marine coast. This allows for the reconstruction of local benthic paleocommunities inhabiting the different sub-environments throughout the Quaternary in this region. Chapter 5 gives examples of the biotic interactions that can be recorded in mollusk shells. Chapter 6 focuses on the faunal distribution, faunistic shifts  and even the extinction of some species during the Argentinean and southern South American Quaternary. Chapter 7 follows the morphometric analyses of selected bivalves well represented in the Quaternary fossil record of the Argentinean coast, and looks at shell shape and size variations through time and the reasons for these changes. Chapters 8 and 9 focus on the use of stable oxygen and carbon isotope analysis of marine shells, since this can provide information on paleotemperatures and productivity; in Chap. 8 this analysis is performed on serial samples of different species from different marine deposits, while in Chap. 9 stable isotopes are treated in conjunction with individual growth in selected fossil specimens of three bivalves. The final chapter, Chap. 10, summarizes the strengths and weaknesses of using mollusk shells, and looks at the main achievements of our work and the gaps which need to be filled in.

# References

Cohen KM, Finney SC, Gibbard PL, Fan JX (2013) The ICS international chronostratigraphic chart. Episodes 36:199–204. http://www.episodes.co.in/contents/2013/september/p199-204.pdf

Gibbard PL, Head MJ, Walker MJC et al (2009) Formal ratification of the quaternary system-period and the pleistocene series-epoch with a base at 2.58 Ma. J Quaternary Sci 25:96–102. doi:http://dx.doi.org/10.1002/jqs.1338

Lowe JJ, Walker MJC (1997) Reconstructing quaternary environments, 2nd edn. Pearson and Prentice Hall, London

Wolff EW (2008) What is the "present"? Quat Sci Rev 26:3023–3024. doi:http://dx.doi.org/10.1016/j.quascirev.2007.10.008

# Chapter 2
# Taphonomy

**Abstract** Taphonomy is the *"science of the laws of burial"* (Efremov 1940); it involves the transition of animal remains from the biosphere to the lithosphere. In this chapter we refer to various taphonomic attributes (e.g., fragmentation and abrasion, among others) by looking at examples in different environments and different taxa; and we explain the advances made in taphonomy by the working group.

**Keywords** Southern South America · Quaternary · Pleistocene · Holocene · Mollusca · Taphonomy · Taphonomic attributes · Taphonomic grades · Fragmentation · Dissolution

## 2.1 Taphonomy as a Cross-Disciplinary Branch

Taphonomy is the *"science of the laws of burial"* (Efremov 1940), *"a branch of paleontology, and almost a branch of ecology"* (Gordillo 2011), and–as it involves the transition of animal remains from the biosphere to the lithosphere–taphonomy is also the study of the death and decay of organisms, including the process of fossilization.

The terms "authochthonous", "parautochthonous" and "allochthonous" proposed by Kidwell et al. (1986) are currently used by many authors to describe the nature of preserved mollusk assemblages. If they are transported, and therefore indicative of the environments in which they finally ended up, they are allochthonous assemblages. If the preserved shells are recovered in situ, and therefore record the environment in which they lived, they are authoctonous assemblages. Finally, if they reflect a situation with locally reworked faunas, but are essentially in situ, they are parautochthonous assemblages.

In this respect, the history of shell burial and exhumation is strongly associated to the taphonomically active zone (TAZ), defined by Aller (1982) as the zone near the sediment–water interface where pore waters are undersaturated with respect to

S. Gordillo et al., *Mollusk Shells as bio-geo-archives*,
South America and the Southern Hemisphere, DOI: 10.1007/978-3-319-03476-8_2,
© The Author(s) 2014

aragonite and calcite, and where most dissolution of carbonate minerals occurs. The length of time a shell remains at the surface and the time it spends buried just below the surface in the TAZ are significant factors in determining whether the shell will become part of a preserved fossil assemblage (Parsons-Hubbard et al. 1999). For futher details on taphonomy of marine shelly faunas see Kidwell and Bosence (1991).

## 2.2  Taphonomic Attributes

The nature of a fossil concentration is defined by taphonomic attributes (i.e., preservational features), and this kind of study allows death assemblages to be interpreted by observing the fossil remains. One way to do this is to analyze different taphonomic attributes according to different taphonomic grades (Fig. 2.1), for each sample and for a target species (i.e., the most common and/or best preserved species). The results are then averaged over the entire sample for comparisons with other sites/regions. Some of the taphonomic attributes most frequently used in mollusk shells are the following:

The *ratio of opposite valves* refers to the number of left and right valves of a particular species in each assemblage, and this feature is useful for evaluating transport from the original community. Another common feature is *fragmentation*, which is associated with the breakage of shells and serves as a proxy for environmental energy. The degree of shell fragmentation tends to be highest in environments with high water turbulence and coarse substrates, such as beaches and tidal channels, due to the impact of other shells, rocks and waves (Parsons and Brett 1991), although it can be influenced by ecological interactions, like shell-breaking predation or bioturbation (Zuschin et al. 2003). As an example, the degree of fragmentation can be estimated following a three-grade scale: whole shell (unbroken), broken shell (up to 50 % broken) and fragment (more than 50 % of the entire shell is absent).

It is also important to analyze the degree of surface alteration, which is generally related to the abrasion, corrosion or bioerosion of skeletons (Parsons and Brett 1991). *Abrasion* of shells occurs when they are exposed to moving particles or when the shells themselves are moved relative to other particles. It is produced by near-shore waves, currents or tidal action, and the most common effect of abrasion on mollusk shells is the loss of surface ornamentation. *Corrosion* on shell surfaces is frequently produced by chemical dissolution. Different skeletons display different solubility in acidic solutions. Calcitic hard parts with high magnesium carbonate content (e.g., echinoderms) are the most soluble, followed by aragonitic and low magnesium calcitic hard parts (Flessa and Brown 1983). Brett and Baird (1986) introduced the term *corrasion* to describe the degradation of shell surfaces when it is difficult to distinguish between abrasion and corrosion. *Wear*, related to abrasive agents, is also used to evaluate surface alteration. *Bioerosion* refers to the alteration of shells through the activity of organisms, usually in search of either food or shelter, and may take the form of boring, rasping, etching, breakage or abrasion of the shell. Another feature, *encrustation*, which refers to

**Fig. 2.1** *Tawera gayi* shells displaying three taphonomic grades for fragmentation, wear, and bioerosion and encrustation. **a** Unbroken shell, **b** Broken shell, **c** Fragmented shell, **d** Shell with ornamented surface, not abraded, **e** Shell with abraded surface, **f** Shell with internal layer exposed, **g** Unbored shell, **h** Bored shell, **i** Shell with incomplete drill-hole, **j** Shell without encrusters, **k** Shell with encrusters on external surface, **l** Shell with encrusters on the internal surface (after Gordillo et al. 2011)

organisms that use shells to live, is a good indicator of the duration of shell exposure at the sediment–water interface (Parsons and Brett 1991); hard-parts are often encrusted after death.

To analyze the percentage of fragmentation, surface alteration or bioerosion over an entire sample, different taphonomic categories are useful for further interpretations.

Finally, *size-sorting* involves the segregation of fossil elements, and reflects prolonged exposure to currents, selective winnowing and transport of shells by currents throughout different hydrodynamic events (Speyer and Brett 1988).

The analysis of ecological features such as mode of life is also useful for taphonomic studies. When considering life position with respect to sediment, mollusks are classified as *epifauna* when they live on a surface such as the sea floor, or on other organisms, and as *infauna* when they live in the substrate, especially when they are buried in a soft sea bottom. However, intergradation between the two categories makes this classification somewhat arbitrary and artificial, so a third intermediate category, *semi-infaunal* (Stanley 1970), can be applied to organisms that live partially buried in the substratum.

## 2.3  Exhuming Clams and Chitons

In southern South America, the shell remains of living and fossil specimens of five bivalves, including mytilids and venerids from the Beagle Channel and the Strait of Magellan, have provided valuable clues to local variations in physical factors such as current speed, wave action and freshwater input along these coasts during the Holocene (Cárdenas and Gordillo 2009; Gordillo et al. 2010, 2011).

According to taphonomic analyses, two types of environment exist on the coasts of Tierra del Fuego: a high energy environment in the Strait of Magellan terraces and a low to moderate energy environment in the Beagle Channel. In the Strait of Magellan, the mytilids (epifauna) show high fragmentation and abrasion, thus implying that they were subject to long exposure on the sea bottom before burial. It is also possible that these taxa experienced shell transport in abrasive sediment due to currents in a high-energy setting such as a foreshore environment, and/or multiple reworking episodes (Speyer and Brett 1991). In the Beagle Channel, fossil assemblages have moderate fragmentation and abrasion, thus indicating a low to moderate energy environment with a dominance of soft bottoms such as sand or small gravel. Venerids (infaunal or semi-infaunal species) are well-preserved in both zones of Tierra del Fuego, and although abrasion and fragmentation in the deposits are moderate, this can be attributed to high bioerosion of their valves caused mainly by boring gastropods and/or encrusting elements on the surface of the shell (Zuschin and Stanton 2001). In this respect, results show a relationship between the levels of bioerosion and fragmentation in the deposits from the Strait of Magellan, i.e., deposits with high bioerosion also show high fragmentation. Nonetheless, in other sites along the Beagle Channel

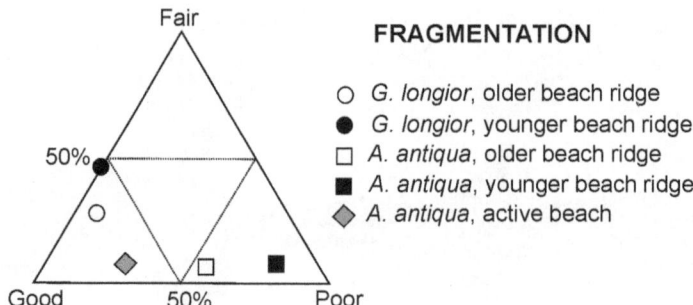

**Fig. 2.2** Distribution of different taphonomic grades for the attribute fragmentation, considered in *Glycymeris longior* and *Ameghinomya antiqua* from Holocene beach ridges and modern *A. antiqua* shells from the present day (active) beach. The graph shows a higher percentage of broken shells in the youngest beach ridge, thus implying different environmental conditions with respect to the older one. *A. antiqua* samples from these deposits are closer to the Poor apex of the *triangle*, thus indicating shells with fractures greater than 50 %

(the Alakush and Ushuaia sites), this relationship does not exist and the high fragmentation of venerids could be associated with different postmortem processes. In particular, venerids showed a higher preservation potential than mytilids, and their infaunal life cycle and fast burial rate make them more reliable specimens to use in further taphonomic analyses.

In Patagonia, Argentina, the coastal area of Puerto Lobos (Chubut, northern Patagonia, 42°00′S 65°4′W) was chosen for the taphonomic analysis (Fig. 2.2) of two common bivalve species from the area (*Glycymeris longior* and *Ameghinomya antiqua*). This was useful for evaluating changes in environmental conditions, such as waves and currents (Boretto et al. 2013). *Glycymeris longior* is typical of the Argentinean Province, located to the north, while *Ameghinomya antiqua* is typical of the Magellan Province, located to the south. However, the two species coexist in Puerto Lobos, which is located between these two malacological provinces. For taphonomic analysis, mollusk shells (N = 268) were collected from the active beach and from two Holocene beach ridges: the older ridge was dated at 3310 ± 90 years BP, while the younger ridge was dated at 750 ± 75 years BP (Bayarsky and Codignotto 1982).

Data was analyzed using the ternary taphograms proposed by Kowalewski et al. (1995), which have a semiquantitative character. The corners of the equilateral triangles are classified as Good (little or no development of a particular attribute), Fair (moderate development of a taphonomic characteristic), or Poor (high development of the attribute). The entire sample can be characterized by the proportion of shells in each of the three categories (Kowalewski et al. 1995). For Fragmentation, the grade Good indicates no fractures; Fair represents less than 50 % broken shells; and Poor is when more than 50 % of the shells are broken. *G. longior* assemblages from the older Holocene deposit are located close to the Good grade, with a tendency towards Fair, thus indicating the preservation of whole shells, or shells that are less than 50 % broken. It also shows a higher percentage of fragmented *G. longior* shells

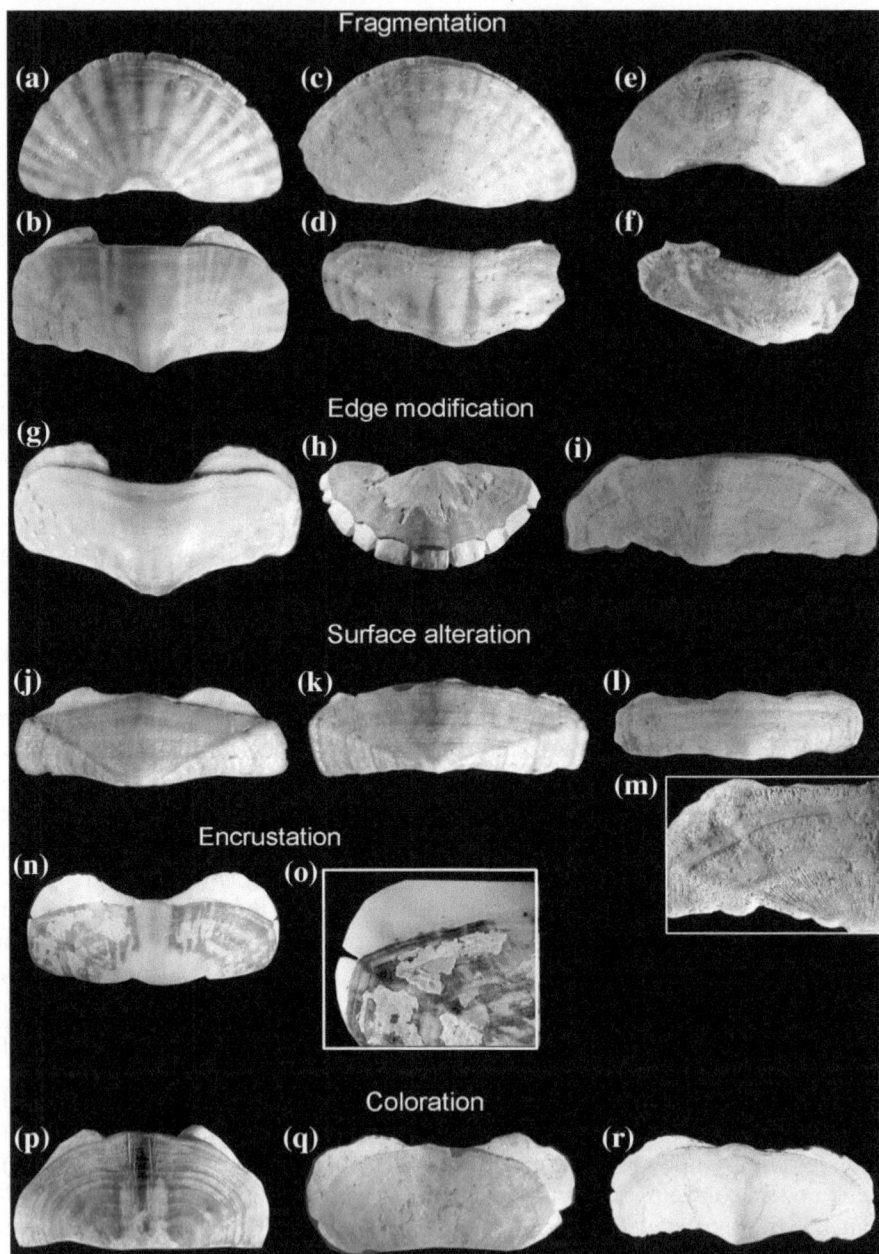

**Fig. 2.3** Taphonomic attributes and grades using chiton plates. Fragmentation: **a, b** Whole plates, **c, d** Broken plates, **e, f** Fragments. Edge modification: **g** Edge without modification, **h** Chipped edge, **i** Polished edge. Surface alteration: **j** Surface without changes, **k** Surface alteration up to 50 %, **l, m** Completely altered surface. Encrustation: **n, o** Intermediate plate with encrusting algae. *Coloration* **p** Plate with original color, **q** Discolored plate, **r** Colorless plate (*white* washed) (after Gordillo 2007)

in the younger deposit, thus implying different environmental conditions. *A. antiqua* samples from these deposits are closer to the Poor apex of the triangle, thus indicating shells with fractures greater than 50 %. As in the previous case, samples from the younger deposit display a higher degree of fracturing than samples collected from the older deposit. *A. antiqua* shells sampled from the modern beach are preserved whole (Good-Poor grade). These trends indicate greater energy in the depositional environment for the youngest beach ridge, since both species are more vulnerable to fracture in relation to those analyzed in the oldest beach ridge. Nevertheless, these results are best interpreted in conjunction with intrinsic properties of resistance to shell breakage (see Chap. 3).

A third example comes from the analysis of taphonomic and paleontological attributes in Holocene chitons (Fig. 2.3). Gordillo (2007) showed that the taphonomic condition of a chiton plate is the result of biological, ecological and environmental factors, and in this study the surface of a high percentage of chiton plates was affected. It was also considered highly probable that the dissolution process is post-depositional and is associated with changes in pH, since several of these deposits are currently associated with brackish or freshwater environments. This situation, coupled with the frequent rainfall and snow in the region, would have led to an acidic environment that favored the dissolution of the carbonates within the plates. However, unlike chitons, dissolution did not significantly alter the bivalves present in the same associations. The reason for this inequality may be linked to differences in the microstructure and the organic content of the bivalve shells and chiton plates. Apparently, the chiton plates have pores associated with an interconnecting network of microtubules (for housing soft tissue with a sensory function), and this raises the proportion of organic content and the flow capacity of interstitial water, thus favoring dissolution. The environmental acidity as a cause of dissolution, and the relationship between microstructure and differential preservation of mollusk shells, has also been considered by other authors (e.g., Glover and Kidwell 1993; Isaji 1993). The exposure of internal layers obtained from raised beach plates could also be associated with some epibionts not preserved in the fossil record (e.g., algae, sponges), which could have acted as agents of bioerosion and facilitators of dissolution. This example with chiton plates shows that they are also suitable for evaluating the post-mortem events that took place up to when they were found within the fossil record.

# References

Aller RC (1982) Carbonate dissolution in nearshore terrigenous muds: the role of physical and biological reworking. J Geol 90:79–95. doi:http://dx.doi.org/10.1086/628652

Bayarsky A, Codignotto JO (1982) Pleistoceno Holoceno marino en Puerto Lobos, Chubut. Rev Asoc Geol Arg 37:91–99

Boretto G, Gordillo S, Cioccale M, Colombo F, Fucks E (2013) Multi-proxy evidence of late quaternary environmental changes in the coastal area of Puerto Lobos (Northern Patagonia, Argentina). Quat Int 305:188–205. doi:http://dx.doi.org/10.1016/j.quaint.2013.02.017

Brett CE, Baird GC (1986) Comparative taphonomy: a key to paleoenvironmental interpretation based on fossil preservation. Palaios 1:207–227. doi:http://dx.doi.org/10.2307/3514686

Cárdenas J, Gordillo S (2009) Paleoenvironmental interpretation of late quaternary molluscan assemblages from southern South America: a taphonomic comparison between the strait of Magellan and the Beagle Channel. Andean Geol 36:81–93. doi:http://dx.doi.org/10.4067/S0718-71062009000100007

Efremov IA (1940) Taphonomy, a new branch of paleontology. Akad Nauk SSSR byull, Seriya Biologicheskaya 3:405–413. doi:http://iae.newmail.ru/science/taph.htm

Flessa KW, Brown TJ (1983) Selective solution of macroinvertebrate calcareous hard parts—a laboratory study. Lethaia 16:193–205. doi:http://dx.doi.org/10.1111/j.1502-3931.1983.tb00654.x

Glover C, Kidwell S (1993) Influence of organic matrix on the post-mortem destruction of molluscan shells. J Geol 101:729–747. doi:http://dx.doi.org/10.1086/648271

Gordillo S (2007) Análisis tafonómico de quitones (Polyplacophora: Mollusca) holocenos de Tierra del Fuego, Argentina. Ameghiniana 44:407–416. doi:http://www.scielo.org.ar/scielo.php?script=sci_arttext&pid=S0002-70142007000200012&lng=es&nrm=iso&tlng=es

Gordillo S (2011) Shell remains of living and fossil clams. JMBA Global Mar Env 14:30–31

Gordillo S, Bayer MS, Martinelli J (2010) Moluscos recientes del Canal Beagle, Tierra del Fuego: un análisis cualitativo y cuantitativo de los ensambles de valvas fósiles y actuales. An Inst Pat 38:95–106. doi:http://www.scielo.cl/scielo.php?script=sci_arttext&pid=S0718-686X2010000200009

Gordillo S, Martinelli J, Cárdenas J, Bayer S (2011) Testing ecological and environmental changes during the last 6000 years: a multiproxy approach based on the bivalve *Tawera gayi* from southern South America. J Mar Biol Ass UK 91:1413–1427. doi:http://dx.doi.org/10.1017/S0025315410002183

Isaji S (1993) Formation of organic sheets in the inner shell layer of *Geloina* (Bivalvia: Corbiculidae): an adaptive response to shell dissolution. Veliger 36:166–173

Kidwell SM, Bosence DWJ (1991) Taphonomy and time-averaging of marine shelly faunas. In: Allison PA, Briggs DE (eds) Taphonomy: releasing the data locked in the fossil record. Topics in geobiology, vol 9. Plenum Press, New York, pp 115–209. doi:http://geosci-webdev.uchicago.edu/pdfs/kidwell/1991KidwellBosenceoptA.pdf

Kidwell SM, Fursich FT, Aigner T (1986) Conceptual framework for the analysis of fossil concentrations. Palaios 1:228–238. doi:http://dx.doi.org/10.2307/3514687

Kowalewski M, Flessa KW, Hallman DP (1995) Ternary taphograms: triangular diagrams applied to taphonomic analysis. Palaios 10:478–483. doi:http://dx.doi.org/10.2307/3515049

Parsons KM, Brett CE (1991) Taphonomic processes and biases in modern marine environments: an actualistic perspective on fossil assemblage preservation. In: Donovan SK (ed) The processes of fossilization. Belhaven Press, London, pp 22–65

Parsons-Hubbard KM, Callender WR, Powell ER, Brett CE, Walker SE, Raymond AL, Staff GE (1999) Rates of burial and disturbance of experimentally-deployed molluscs: implications for preservation potential. Palaios 14:337–351. doi:http://dx.doi.org/10.2307/3515461

Speyer SE, Brett CE (1988) Taphofacies models for epeiric sea environments-middle Paleozoic examples. Palaeogeogr Palaeocl 3:225–262. doi:http://dx.doi.org/10.1016/0031-0182(88)90098-3

Speyer SE, Brett CE (1991) Taphofacies controls background and episodic processes in fossil assemblage preservation. In: Allison PA, Briggs DE (eds) Taphonomy: Releasing data locked in the fossil record, Topics in Geobiology 9. Plenum Press, New York, pp 501–545

Stanley SM (1970) Relation of shell form to life habits of the Bivalvia (mollusca). Boulder, Geol Soc Am Inc. doi:http://dx.doi.org/10.1130/MEM125-p1

Zuschin M, Stanton RJ (2001) Paleocommunity reconstruction from shell beds: a case study from the main glauconite bed, eocene, Texas Palaios 17:602–614

Zuschin M, Stachowitsch M, Stanton RJ Jr (2003) Patterns and processes of shell fragmentation in modern and ancient marine environments. Earth-Sci Rev 63:33–82. doi:http://dx.doi.org/10.1016/S0012-8252(03)00014-X

# Chapter 3
# Shell Microstructure and Shell Architecture

**Abstract** This chapter is focused on the analysis of the shell microstructure of different taxa and on how this information can be used for paleoenvironmental interpretations. A physic-chemical analysis on Modern, Holocene and Pleistocene shells of the purple clam *Amiantis purpurata* helps discern the structural changes during early diagenesis. In addition, the analyses of the microstructure of two other bivalves (*Glycymeris longior* and *Ameghinomya antiqua*) from the same region explain the differences in the degree of fragmentation in both species as a result of different structural features. Finally, cathodoluminiscence applied to *Tawera gayi* provides information on the skeletal growth cycles that is useful for evaluating changes.

**Keywords** Southern South America · Quaternary · Pleistocene · Holocene · Mollusca · Shell microstructure · Mechanical resistance · Early diagenesis

Shell microstructure refers to the arrangement of basic microstructural units, such as tablets, rods and blades, in a shell layer. Shell architecture deals with the larger aspect of the shell microstructure, such as the orientation of the largest units of shell microstructure with respect to the shell form (Carter 1980).

## 3.1 Inorganic-Organic Biocomposites

As seen in Chap. 2, mollusk shells have been used in comparative taphonomic studies because of their excellent potential for preservation. However, their resistant hard parts are subject to physical, chemical and biological agents or processes that can destroy these shells before and after burial e.g., (Lawrence 1968).

Bivalve shells are predominately composed of $CaCO_3$, in other words, calcite, variable proportions of aragonite, even vaterite, as well as organic polymers (Hare and Abelson 1965; Rhoads and Lutz 1980; Nehrke et al. 2012). As in other biomineralized exoskeletons, the orientation and growth of $CaCO_3$ crystals are strongly controlled by the organic matrices (which constitute about 1–5 wt% of the shell) forming compartments in which mineralization takes place. Most of the organic components are intercrystalline, with a smaller portion located within the crystal structure of calcium carbonate. This mixture, on a very fine scale of organic and bioprecipitated $CaCO_3$, modifies diagenetic processes and patterns in comparison with non-biogenic mineral features (Perrin and Smith 2007).

Mollusk shells can be considered inorganic-organic biocomposites, with excellent mechanical performance compared to non-biogenic material (Chateigner et al. 2010). Even though aragonite provides high mechanical strength to the valve (Chateigner et al. 2000), under the environmental conditions found on the Earth's surface, this phase is metastable, and is more susceptible to dissolution and recrystallization than calcite. In other words, during diagenesis, the alteration of aragonite skeletons commonly results in mineralogical and structural changes, as well as compression and postdepositional cementation (Brand 1989). For this reason, aragonitic fossils are usually poorly preserved in the geological record (Powell and Kowalewski 2002; Cherns et al. 2011). One exception is for the Cenozoic A-seas (aragonitic seas), in which a positive bias favoring aragonitic bivalves is recognized (De Renzi and Ros 2002).

This relative instability of aragonite when exposed to diagenesis introduces a bias in the fossil record, thus affecting its quality for paleoecological and paleo-biological studies (Fernandez Lopez 2000; Cherns and Wright 2009). The detection of subtle postdepositional changes therefore becomes of utmost importance, especially in fossils that appear to retain their primary mineralogy when conventional screening techniques (such as X-ray diffraction) are used.

## 3.2  Chemical and Physico-Mechanical Changes from Fossil to Modern Shells

In a recent study, Bayer et al. (2013) described the chemical (trace element), textural, and physical–mechanical transformation that took place in shells of an aragonitic bivalve, the *Amiantis purpurata* venerid, during early diagenesis in a period of time that exceeds 100,000 years (i.e., from the Late Pleistocene up to the present day). The advantage of considering the same species from different outcrops of the same region is the elimination of interspecific variations associated with intrinsic factors (shell microstructure) and different environmental conditions that can occur when comparing shells of the same species collected from different regions.

Bayer et al. (2013) found that mineralogy remains constant, with aragonite as the only crystalline phase throughout the entire examined time interval, but texture (as revealed by XRD, SEM and optical microscopy; Fig. 3.1) is modified. The Pleistocene valve has more grains in a random distribution, compared with the twinning pattern evident in Modern and Holocene shells (Fig. 3.1). The approximate constant value of crystallite size suggests that the dissolved $CaCO_3$ does not

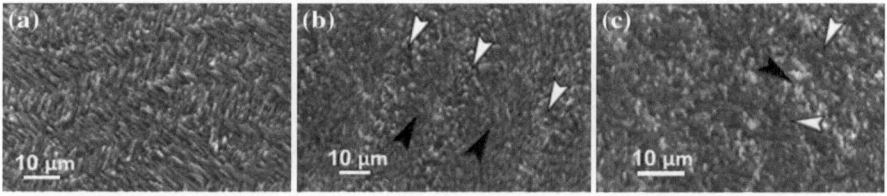

**Fig. 3.1** SEM images (secondary electron mode) of *Amiantis purpurata* shells. **a** Details of the growth bands of a Modern shell, showing platy aragonite crystals defining a crossed-lamellar microstructure. **b** Growth bands from a Holocene shell that can still be observed. Some bands (marked with *white arrows*) have been replaced by a granular aggregate of aragonite crystals, whereas others (*black arrows*) still display the crossed-lamellar microstructure. **c** View of a Pleistocene shell where no growth bands can be discerned, showing a mixture of biogenic platy crystals (*white arrows*) with equant grains of aragonite of diagenetic origin (*black arrow*) (Modified from Bayer et al. 2013)

precipitate in crystallographic continuity with the preexisting crystals, as this would lead to a larger crystallite size (and narrower diffraction peaks) and a sharpening of the twinning pattern. A possible explanation which remains to be tested is that aragonite cement occupies voids left by degraded organic matrix, as found by Webb et al. (2007).

A comparison of the chemical composition of *A. purpurata* valves of different ages (Bayer et al. 2013) shows trends of decreasing Na, Sr and (to a lesser degree) Mg with increasing age. The authors correlate these variations with the dissolution and reprecipitation of aragonite through a thin film of solutions of meteoric origin in small-scale, semiclosed microenvironments (a vadose environment).

In terms of mechanical properties, a shell's resistance to fracture is characterized by the measurement of its strength (Zuschin and Stanton 2001; Yang et al. 2011). The mechanical properties of the valves, such as the ability to transfer charges between adjacent layers of aragonite (Liang et al. 2008), are associated with the structure and functions of the biological matrix, which also promotes the formation of crystalline layers (Rhoads and Lutz 1980). Bayer et al. (2013) found that Holocene and Pleistocene *A. purpurata* shells have a higher Vickers hardness and a more fragmented area than the Modern shells (Fig. 3.2). These differences indicate that over a period of less than 5,000 years the valves of *A. purpurata* have become harder but more brittle. These changes are also attributed to postdepositional modifications by dissolution-recrystallization processes mediated by a thin film of water in a vadose environment.

Microstructural adjustments are more sluggish than chemical modifications produced by diagenetic processes, whereas microhardness rapidly reaches high values, probably due to the early degradation of organic shell compounds.

In conclusion, this study shows that chemical and physico-mechanical changes in mollusk shells start early (at least before a shell reached the age of 5,000 year BP), increase with age and most probably occurred as a consequence of the degradation of the skeletal organic matrix.

**Fig. 3.2** Changes in
mechanical resistance
measured as microhardness
of Modern, Holocene and
Pleistocene samples of
*Amiantis purpurata* (after
Bayer et al. 2013)

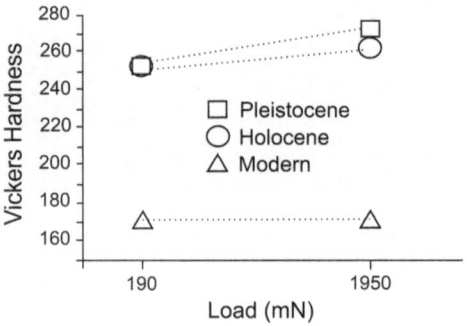

## 3.3 The Intrinsic Properties of Taxa Lead to Differential Behavior Under Environmental Conditions

Boretto et al. (2013) described the microstructure of two bivalves (*Glycymeris longior* and *Ameghinomya antiqua*) from Puerto Lobos in northern Patagonia (Argentina), and provided arguments to explain the differences in the degree of fragmentation in both species as a result of different structural features. The associations of shells analyzed in these Holocene beach ridges can be classified as allochthonous (Kidwell and Bosence 1991), since they originated in the intertidal and subtidal zones (production areas) and were transported to the supralittoral area (settling zone). Taphonomic analyses, which evaluate the energy of the processes involved in the formation of these assemblages, indicate that *A. antiqua* and *G. longior* shells were drawn together during high energy storm events, and that these bivalve assemblages were affected by the same transport conditions and physicochemical characteristics as the depositional environment. Previous X-ray diffraction studies performed on samples of both taxa (Bolmaro et al. 2006; Boretto et al. 2013) indicate their aragonitic mineralogical composition. On the one hand, *A. antiqua* (Fig. 3.3a–c) is characterized by a dominant prismatic microstructure with two aragonitic layers: an outer prismatic layer and a homogenous inner layer, although sometimes a thinner, crossed-lamellar third layer can be observed (Carter 1990). This species shares the same attributes described for *Protothaca thaca* from Chile (Lazareth et al. 2006). On the other hand, in *G. longior* (Fig. 3.3d–f), two major shell layers can be distinguished: the inner and the outer layers, separated by the thin miostracal and middle layers. Additionally, the outer shell layer can be further subdivided into a middle layer and an outer layer. In this respect, *G. longior* shares the same attributes as *G. glycymeris* (Rogalla and Amler 2007). Boretto et al. (2013) described *G. longior* as composed of an inner shell layer which has a complex crossed lamellar structure, with interdigitating first order lamellar of lenticular shape that extend normal to the shell, and an area of homogeneous microstructure close to the apex. The miostracal layer of a fairly constant thickness (between 50 and 100 mm) and a prismatic structure is then developed, composed of elongated prisms with the long

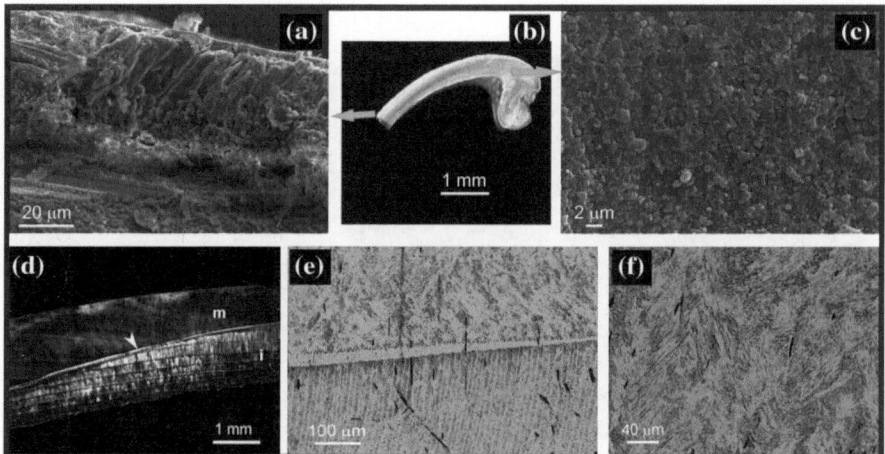

**Fig. 3.3** Microstructural SEM images of *Ameghinomya antiqua* (**a–c**) versus *Glycymeris longior* (**d–f**). **a** Outer prismatic layer. **b** Cross-section from the umbo along the growth axis, showing the outer and inner layers. **c** Homogenous *inner* layer. **d** Cross-section of the shell showing the cloudy inner "i" and middle "m" layers which contrast with the *white arrow* point to the clear miostracal layer. **e** Contact of the miostracal layer with the inner and middle layers; note that the miostracal layer is homogeneous and lacks the *dark* spots which are interpreted as zones rich in organic material. **f** Close-up of the inner layer, showing complex *crossed* lamellar structure, first and second order lamellar (these defining a chevron pattern) (after Boretto et al. 2013)

axes oriented perpendicular to the layer margin. Below this, a crossed lamellar structure defines the middle layer, and finally these lamellar become thicker, torted and anastomosing, thus transitionally defining the outer zone. These differences in microstructure explain the different behavior of both species in relation to the taphonomic attributes analyzed in Chap. 2. Studies on the mechanical strength at break indicate that the crossed lamellar microstructure has a better performance in the elastic range, with a higher effective Young's modulus than the prismatic structure (Bolmaro et al. 2006). This explains the *G. longior* overall shell preservation as "whole and some broken", due to the complex cross lamellar microstructure, compared with the samples of *A. antiqua* from the Puerto Lobos site, which have a high degree of fragmentation.

# 3.4 Cathodoluminiscence Applied to Biogenic Carbonates

Although the X-ray examination of shells provides information on their mineralogical composition, there are other types of analysis, such as cathodoluminiscence (CL) applied to recent benthic biogenic carbonates (e.g., mollusk shells), through which information on the microstructure of shells and their growth can be obtained (Barbin 1992; Barbin and Gaspard 1995; Gordillo et al. 2011).

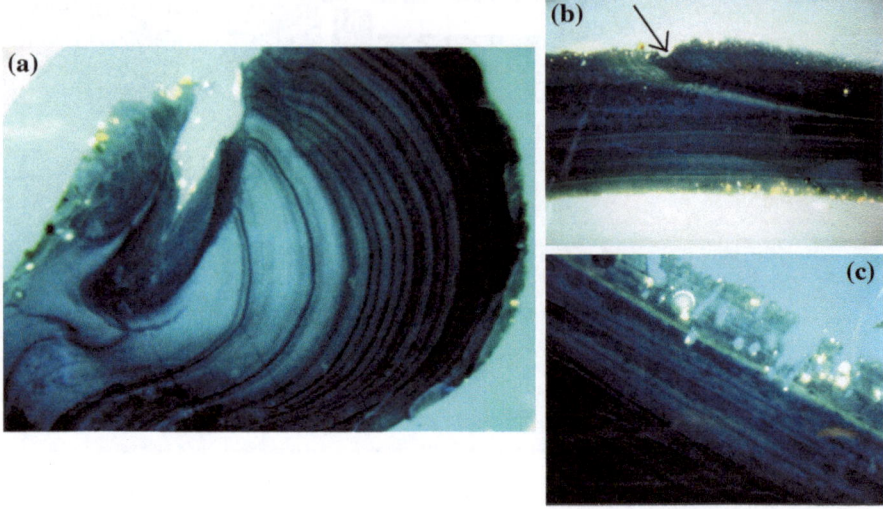

Fig. 3.4  View under cathodoluminiscence (CL) of sections of modern (**a**) and fossil (**b-c**) *Tawera gayi* shells from southern South America, showing a well-defined, almost concentric pattern of CL lines. Luminiscent bands border the winter (*dark*) growth rings; **c** high magnification (10x) of (**b**). A different luminescence (*light, bright yellow* luminescence) affecting *outer* and *inner* shell surfaces is interpreted as a bioeroded surface caused by external factors (i.e., bacteria and microboring organisms), but not produced by the mollusk biomineralization process

Gordillo et al. (2011) observed that under CL-microscopy, modern and fossil shells of the venerid *Tawera gayi* exhibit a well-defined pattern, with parallel spaced CL lines (Fig. 3.4). This zonation reflects the cycles of skeletal growth and luminescence intensity typical of aragonitic shells, and may be related to the alternating amount of manganese present in the aragonite (Barbin 1992). A rapid growth rate during the earlier life stages of *T. gayi* (Fig. 3.4a), and CL lines that terminate in an external growth line (Fig. 3.4b), as well as the regular repetition of CL with outlines approaching the shape of internal structures, indicate that these lines are related to the growth dynamics of the shell (see discussion in Tomasovych and Farkas 2005). The aragonitic *T. gayi* shells give a weak blue-green luminescence (probably due to a slower growth rate) alternating with dark areas associated with periods of a different growth rate (or a cessation of growth). In addition, a different luminescence (bright, light yellow luminescence) affecting outer and inner shell surfaces is interpreted as a bioeroded surface caused by external factors (i.e., bacteria and microboring organisms), but not produced by the mollusk biomineralization process (Fig. 3.4c). Although the data presented on the shell structure of *T. gayi* under CL is not enough to explain the true reasons behind the differences or changes among shells, it does indicate that CL lines correspond to zones recording changes in growth rate. Thus, the analysis of CL lines in this species can provide another important tool for the evaluation of *T. gayi* growth rates, in addition to external growth rates, isotopes and trace elements, since CL lines in bivalves are correlated with periods of slow growth, such as winter, spawning seasons or environmental disturbance (Barbin 1992; Barbin and Gaspar 1995). A systematic examination of CL line pattern in *T. gayi* can be useful for adding to our knowledge of changes during the Holocene.

# References

Barbin V (1992) Fluctuation in shell composition in *Nautilus* (Cephaolopoda, Mollusca): evidence from cathodoluminiscence. Lethaia 25:391–400. doi:http://dx.doi.org/10.1111/j.1502-3931.1992.tb01642.x

Barbin V, Gaspard D (1995) Cathodoluminiscence of recent articulate brachiopod shells. Implications for growth stages and diagenesis evaluation. Geobios 18:39–45. doi:10.1007/978-3-662-04086-7_12

Bayer MS, Colombo F, De Vincentis NS, Duarte GA, Bolmaro R, Gordillo S (2013) Cryptic diagenetic changes in Quaternary aragonitic shells: a textural, crystallographic, and trace element study of *Amiantis purpurata* from Patagonia Argentina. Palaios 28:438–451. doi:http://dx.doi.org/10.2110/palo.2012.p12-111r

Bolmaro RE, Romano Trigueros P, Zaefferer S (2006) Estudio de la resistencia mecánica y la textura de los caparazones mineralizados de bivalvos. In: Actas 17th congresso brasileiro de engenharia e ciência dos materiais. Foz do Iguaçu, Brasil. doi:http://www.materiales-sam.org.ar/sitio/biblioteca/CONAMET-SAM2006/docs/o4.pdf

Boretto G, Gordillo S, Cioccale M, Colombo F, Fucks E (2013) Multi-proxy evidence of late quaternary environmental changes in the coastal area of Puerto Lobos (Northern Patagonia, Argentina). Quatern Int 305:188–205. doi:http://dx.doi.org/10.1016/j.quaint.2013.02.017

Brand U (1989) Aragonite-calcite transformation based on Pennsylvanian mollusks. Geol Soc Am Bull 101:377–390. doi:http://dx.doi.org/10.1130/0016-7606(1989)101<0377:ACTBOP>2.3.CO;2

Carter JG (1980) Guide to bivalve shell microstructures. In: Rhoads DC, Lutz RA (eds) Skeletal growth of aquatic organisms. Plenum, New York

Carter JG (1990) Evolutionary significance of shell microstructure in the Paleotaxodonta, Pteriomorphia and Isofilibranchia (Bivalvia: Mollusca). In: Carter JG (ed) Skeletal biomineralization: patterns, processes, and evolutionary trends. Van Nostrand Reinhold, New York, pp 135–296

Chateigner D, Hedegaard C, Wenk HR (2000) Mollusc shell microstructures and crystallographic textures. J Struct Geol 22:1723–1735

Chateigner D, Ouhenia S, Krauss C, Belkhir M, Morales M (2010) Structural distortion of biogenic aragonite in strongly textured mollusk shell layers. Nucl Instrum Methods 268:341–345. doi:http://dx.doi.org/10.1016/j.nimb.2009.07.007

Cherns L, Wright VP (2009) Quantifying the impacts of early diagenetic aragonite dissolution on the fossil record. Palaios 24:756–771. doi:http://dx.doi.org/10.2110/palo.2008.p08-134r

Cherns L, Wheeley JR, Wright VP (2011) Taphonomic bias in shelly faunas through time: early aragonitic dissolution and its implications for the fossil record. Taphonomy Top Geobiol 32:79–105. doi:http://dx.doi.org/10.1007/978-90-481-8643-3_3

De Renzi M, Ros S (2002) How do factors affecting preservation influence our perception of rates of evolution and extinction? The case of bivalve diversity during the Phanerozoic. In: De Renzi MPA, Belinchón MD (eds) Current topics on taphonomy and fossilization. Collecció Encontres, Valencia, pp 77–88

Fernández López SR (2000) Tafonomía, Departamento de Paleontología, Universidad Complutense de Madrid, Madrid, doi:http://eprints.ucm.es/22003/1/087_00_Temas_Tafonomia.pdf

Gordillo S, Martinelli J, Cárdenas J, Bayer S (2011) Testing ecological and environmental changes during the last 6,000 years: a multiproxy approach based on the bivalve *Tawera gayi* from southern South America. J Mar Biol Ass UK 91:1413–1427. doi:http://dx.doi.org/10. 1017/S0025315410002183

Hare PE, Abelson PH (1965) Amino acid composition of some calcite proteins. Carnegie Institution, Washington, pp 223–232

Kidwell SM, Bosence D (1991) Taphonomy and Time-averaging of marine shelly faunas. In: Allison PA, Briggs DEG (eds) Taphonomy. Plenum Press, New York, pp 115–209. doi:http:// geosci-webdev.uchicago.edu/pdfs/kidwell/1991KidwellBosenceoptA.pdf

Lazareth CE, Lasne G, Ortlieb L (2006) Growth anomalies in *Protothaca thaca* (Mollusca, Veneridae) shells: markers of ENSO conditions? Climate Res 30:263–269. doi:http://dx.doi. org/10.3354/cr030263

Lawrence DR (1968). Taphonomy and information losses in fossil communities. Geol Soc Am Bull 79:1315–1330. doi:http://dx.doi.org/10.1130/0016-7606(1968)79[1315:TAILIF]2.0.CO;2

Liang Y, Zhao J, Wang L, Li F (2008) The relationship between mechanical properties and crossed-lamellar structure of mollusk shells. Mater Sci Eng A 483–484:309–312. doi:http:// dx.doi.org/10.1016/j.msea.2006.09.156

Nehrke G, Poigner H, Wilhelms-Dick D, Brey T, Abele D (2012) Coexistence of three calcium carbonate polymorphs in the shell of the Antarctic clam *Laternula elliptica*. Geochem Geophys Geosyst 13(5):Q05014. doi:10.1029/2011GC003996

Perrin C, Smith DC (2007) Earliest steps of diagenesis in living Scleractinian corals: evidence from ultrastructural pattern and Raman spectroscopy. J Sediment Res 77:495–507. doi:http:// dx.doi.org/10.2110/jsr.2007.051

Powell MG, Kowalewski M (2002) Increase in evenness and sampled alpha diversity through the Phanerozoic: Comparison of early Paleozoic and Cenozoic marine fossil assemblages. Geology 30:331. doi:http://dx.doi.org/10.1130/0091-7613(2002)030<0331:IIEASA>2.0.CO;2

Rhoads DC, Lutz RA (1980) Skeletal growth of aquatic organisms. Biological records of environmental change. Plenum Press, New York

Rogalla NS, Amler MRW (2007) Statistic approach on taphonomic phenomena in shells of *Glycymeris glycymeris* (Bivalvia: Glycymeridae) and its significance in the fossil record. Paläontol Z 81:334–355

Tomašových A, Farkaš J (2005) Cathodoluminiscence of Late Triassic terebratulid brachiopods: implications for growth patterns. Palaeogeogr Palaeoclimatol 216:215–233 doi:http://dx.doi. org/10.1016/j.palaeo.2004.11.010

Webb GE, Price GJ, Nothdurft LD, Deer L, Rintoul L (2007) Cryptic meteoric diagenesis in freshwater bivalves: Implications for radiocarbon dating. Geology 35:803–806 doi:http://dx. doi.org/10.1130/G23823A.1

Yang W, Kashani N, Li X-W, Zhang G-P, Meyers MA (2011) Structural characterization and mechanical behavior of a bivalve shell (*Saxidomus purpuratus*). Mat Sci Eng C 31:724–729. doi:http://dx.doi.org/10.1016/j.msec.2010.10.003

Zuschin M, Stanton RJ Jr (2001) Experimental measurement of shell strength and its taphonomic interpretation. Palaios 16:161–170. doi:http://dx.doi.org/10.1669/0883-1351(2001)016<0161: EMOSSA>2.0.CO;2

# Chapter 4
# Paleoecology

**Abstract** Mollusk shells are the most common remains in Quaternary marine deposits throughout the Argentinean coastline and southern South America. They are well preserved and, despite the taphonomic bias (i.e., the loss of soft body taxa and post burial processes), Quaternary mollusk assemblages retain useful information about the life habits and habitats of the marine benthos from which they are derived. This chapter is centered on the analysis of the taxa composition of preserved fauna from different Pleistocene and Holocene marine outcrops, located mainly along the maritime Argentinean coast, in order to reconstruct local benthic paleocommunities inhabiting the different sub-environments throughout the Quaternary in this region.

**Keywords** Southern South America · Quaternary · Pleistocene · Holocene · Mollusca · Taphonomy · Paleoecology · Taxa composition · Guild structure

## 4.1 From Dead Remains to Past Communities

Mollusk shells are the most common remains found in Quaternary marine deposits throughout the Argentinean coastline and southern South America (Feruglio 1950; Gordillo 1998). They are well preserved and, despite the taphonomic bias (i.e., the loss of soft body taxa and post burial processes), Quaternary mollusk assemblages retain useful information about the life habits and habitats of the marine benthos from which they are derived (Aitken 1990). Previous available information has shown that mollusks from the southern tip of South America provide a key for the reconstruction of paleocommunities and the evaluation of changes in faunal composition during the Holocene (Gordillo et al. 2005).

In order to reconstruct paleocommunities from fossil shell remains, the first step is the analysis of taxa composition and abundance, coupled with the analysis of taphonomic attributes (Chap. 2). In this respect, a death assemblage is not

S. Gordillo et al., *Mollusk Shells as bio-geo-archives*,
South America and the Southern Hemisphere, DOI: 10.1007/978-3-319-03476-8_4,
© The Author(s) 2014

biologically equivalent to a census of a living community, but instead sums up the dead-shell input (minus shell destruction) over a longer period, thus permitting the accrual of time-averaged species richness (Kidwell 2002). Kidwell (2001) examined the preservation of species abundance in marine death assemblages and noted that species which are dominant in a single live census also dominate the local death assemblage, and species that are rare or not sampled alive are also rare in death assemblages. In southern South America these studies are still emerging, but Archuby et al. (2011) presented preliminary data on the degree of correspondence between living communities and modern death assemblages along the Patagonian Atlantic coast, and mentioned taphonomic processes that affect fidelity.

It can therefore be seen that the fossil record largely portrays the modern faunas from which it is derived. To identify Quaternary mollusks, much information comes from living species, although this is not always complete and depends on the information available for each particular region.

## 4.2 Guild Structure

The ecological characterization of the Quaternary fossil taxa is based on their living representatives. It includes life habit (mode of life) and feeding-type. This data is commonly used to reconstruct the structure of the mollusk assemblages represented at each individual site.

When counting taxa in order to estimate relative abundance, in the case of bivalves it is important to distinguish between the right and left valves, to prevent an individual from being counted twice. In the same way, as chitons are composed of 8 plates, to count individual chitons, the total number of plates should be divided by 8.

*Mode of life*: This considers the life position with respect to the sediment; mollusks are classified as *epifauna*, *infauna*, and *semi-infauna*, as defined in Chap. 2.

*Feeding-type*: mollusks have been described as suspension feeders, deposit feeders, browsers and carnivores. *Deposit feeders* acquire their nutrition from the sediments they inhabit, whereas *suspension feeders* collect food particles suspended above sediments; *browsers* are *hervibores* that encrust vegetation from the surface of substrates they attach to, and *carnivores* feed on dead or decaying as well as active prey.

*Guild structure*: A guild is a group of species that exploit the same kind of environmental resources in a similar way; the term brings together species that overlap significantly in their niche requirements, without considering their taxonomic position (Root 1967). The purpose of guild analysis is to examine the habitat structure of a community as it functioned in some place at some time (Fig. 4.1).

**Fig. 4.1** Guild structure of Quaternary mollusk assemblages from the San Matías Gulf, based on life habit and feeding mode of the fossil fauna recovered

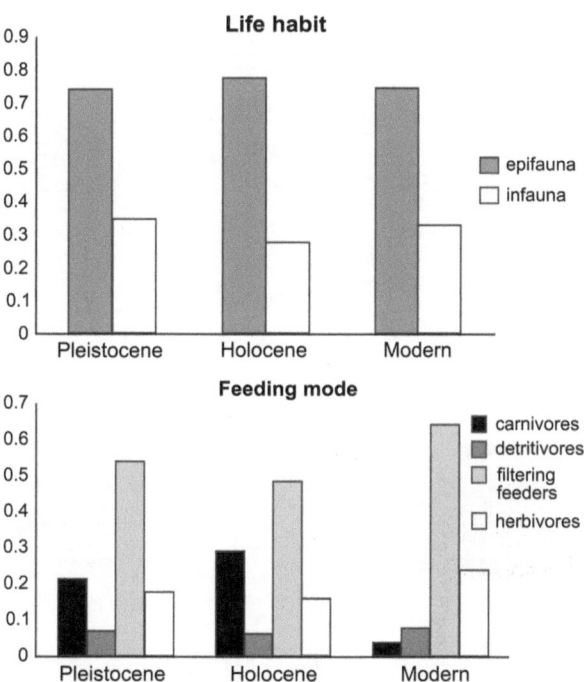

## 4.3 Reconstructing Local Benthic Paleocommunities

During the Quaternary, the southern tip of South America was affected by several glaciations, which might have excluded much of the benthic marine fauna inhabiting this region, and as a consequence the connection between the Atlantic and the Pacific Oceans was broken. In this context, fossil marine mollusks recovered from interglacial (Pleistocene) and postglacial (Holocene) Quaternary deposits in Tierra del Fuego provide a clue for the reconstruction of paleocommunities and the evaluation of changes in faunal composition over time.

Quaternary mollusk assemblages from Tierra del Fuego represent typical shallow benthic paleocommunities which developed during different stages within this period, and variations in faunal composition would mainly be related to differences in substratum types, water depth and sedimentation rates.

For the Beagle Channel, at least four different local communities (three infaunal paleocommunities and one epifaunal paleocommunity) have been recognized (Gordillo 1999; Gordillo et al. 2008).

An epifaunal *Zygochlamys* paleocommunity (ca. 8,000–7,000 BP; Fig. 4.2a) almost completely composed of epifaunal suspension feeders was the most diverse. The long, fragile pectinid shells suggest a quiet-water environment and firm ground substrates more suitable for this group, and the dominance of suspension feeders shows that marine conditions were fully established around 7,500 years ago.

(a)                                      (b)

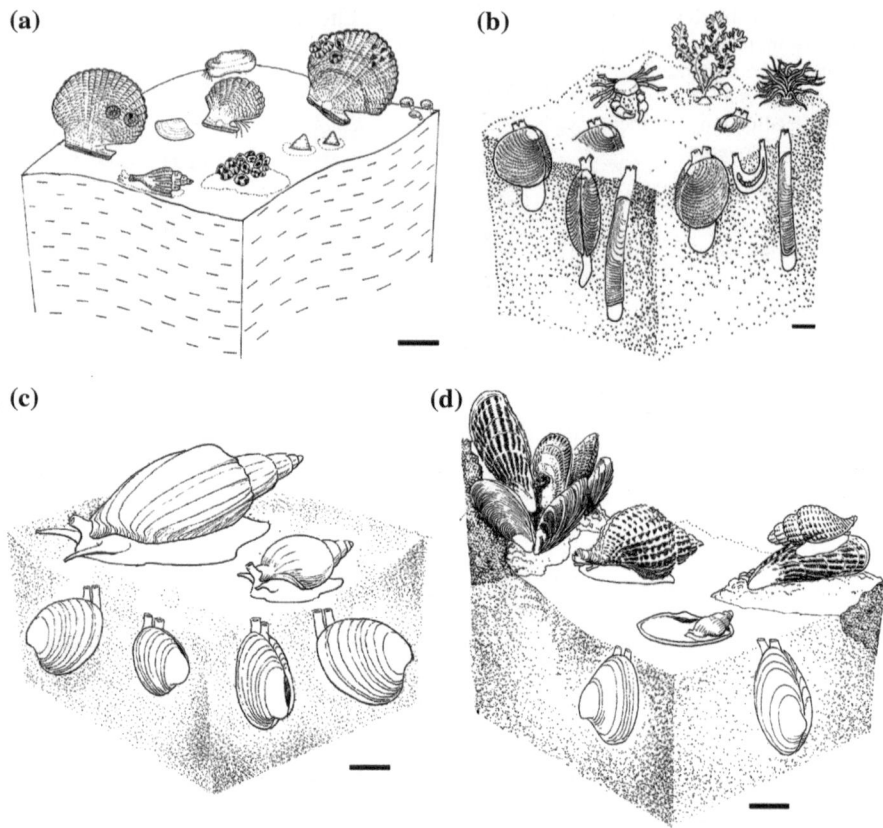

(c)                                      (d)

**Fig. 4.2** Reconstruction of local benthic paleocommunities typical of shallow marine environments in Tierra del Fuego during the Pleistocene-Holocene interval. **a** An epifaunal *Zygochlamys* community developed during the early Holocene in the Beagle channel. **b** An infaunal *Ameghinomya-Ensis* paleocommunity developed in the same channel during the mid-Holocene. **c** A *Retrotapes* infaunal paleocommunity developed during the Pleistocene along the northeastern sector (Atlantic coast) of Tierra del Fuego. **d** A typical paleocommunity of mussels with mixed elements developed in the same area as **c** during the Holocene

Other infaunal local paleocommunities from the same region developed during a period of climatic ameliorization (ca. 4,500–4,000 BP), and are represented by extensive beds of articulated venerids and myoids, many of them even in life position. At least two local benthic communities coexisted and have been described for this region (Gordillo 1999): the *Tawera* paleocommunity and the *Ameghinomya-Hiatella* paleocommunity. A third infaunal paleocommunity (Fig. 4.2b), the *Ameghinomya-Ensis* community (Gordillo 1999; Gordillo et al. 2008), has been described for Golondrina Bay, on the Beagle Channel, and given a radiocarbon age of ca. 7,000–6,000 years BP. This was formed during a phase of climatic deterioration, prior to ca. 5,500 BP, when paleotemperatures and paleosalinities reached their maximum values for the last 8,000 years (Lamy et al. 2002).

For the Atlantic coast, Gordillo and Isla (2011) described two other different local benthic paleocommunities. A *Retrotapes* dominated assemblage (Pleistocene; Fig. 4.2c) corresponds to an infralittoral community dominated by one main infaunal element: the venerid *Retrotapes*, with secondary taxa including other suspension feeders (*Mulinia* and *Mytilus*) and different predatory gastropods (*Trophon, Buccinanops, Odontocymbiola*). Considering the ecological requirements of these taxa and the available paleoenvironmental data, the original community might have developed associated to coarse sand bottoms in the infralittoral zone. The Holocene *Mytilus* intermixed assemblage (Fig. 4.2d) is more diverse, and dominated by sessile suspension feeder epifauna (mytilids) intermixed with some infaunal burrower elements (*Mulinia*), thus suggesting areas with soft substrates suitable for burrower clams. *M. chilensis* lives bysally attached to hard bottoms, forming clusters associated with other species (e.g., *P. purpuratus*, *A. atra*). Other epifaunal elements present are different predatory gastropods (*T. geversianus, Xymenopsis muriciformis, Acanthina monodon*), the buccinid *Pareuthria plumbea* and a variety of limpets (e.g., *Nacella* spp., *Pachysiphonaria*). These taxa are typical of tidal flats and areas more exposed to highly unstable conditions with longer episodes of exposure.

## 4.4  Evaluating Local Faunistic Changes from Mollusk Assemblages

Quaternary mollusk assemblages from Tierra del Fuego represent the typical shallow benthic paleocommunities which developed during different stages within this period, and variations in faunal composition would mainly be related to differences in substratum types, water depth and sedimentation rates.

For the Atlantic coast, Gordillo and Isla (2011) found faunal changes during the Pleistocene-Holocene interval, with a higher proportion of epifaunal elements in the Holocene than in the Pleistocene (Fig. 4.3). The authors explain these changes, and the local disappearance of *Retrotapes* in post-glacial benthic communities in the area, on the basis of regional and local causes, most probably associated to substrate changes, and not through a global trend related to large-scale patterns. In this respect, it is plausible that since Pleistocene transgressions represent higher sea-levels, the ancient bays were deeper and contained a dominance of suspended-feeder specimens in a rather stable soft bottom. As the Holocene sea-level has only fluctuated a few meters, the deposits are dominated by gravel beaches and shallow bays. In this sense, present benthic communities are dominated by epifaunal specimens with suspended feeders tolerant to high-sediment concentrations.

Information on Pleistocene species that inhabited the Beagle Channel before the Last Glacial Maximum (ca. 20,000–18,000 years BP) comes from a paleontological site (Corrales Viejos) located on Navarino Island and described by Gordillo et al. (2010). It is interesting to compare the fauna recovered from this

**Fig. 4.3**  Percentage of
epifaunal and infaunal taxa in
sites of different ages located
along the Atlantic coast of
Tierra del Fuego. An increase
in epifaunal taxa is noted
from older to younger
deposits

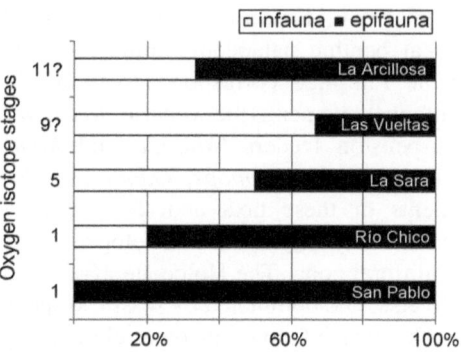

paleontological site with the living fauna in the area, as there are notable differences: the Pleistocene macrofauna is dominated by cirripeds, small nesting bivalves and small muricid gastropods, while the fauna that lives on the adjacent beach has a different living local community, dominated by huge suspension feeder bivalves (mytilids and cirripeds).

In San Blas Bay, southern Buenos Aires Province (Argentina), Charó et al. (2013) also found differences between Pleistocene, Holocene and Modern mollusk assemblages, and the same happened in the San Matías Gulf, northern Río Negro Province (Fig. 4.4). For this latter region, 19 sites (6 Pleistocene, 6 Holocene and 7 Modern) with a total of 42 species (20 bivalves and 22 gastropods) were analyzed. For MIS7, 11 species were recorded and the most abundant were the bivalves *Glycymeris purpurata* and the gastropod *Tegula atra*. For  MIS5e a remarkable diversification of species took place and 22 species were recovered (11 gastropods and 11 bivalves); the most abundant was the bivalve *Brachidontes rodriguezi*, in conjunction with  *Glycymeris longior* and  *Amiantis purpurata*, and the gastropods *Heleobia australis*, *Olivancillaria carcellesi* and *Olivancillaria urceus*. During the MIS1, a similar number of species (23 species, 11 bivalves and 12 gastropods) was recovered, but *Amiantis purpurata* was dominant, in conjunction with the gastropods *Tegula patagonica*, *Buccinanops cochlidium*, *Crepidula* sp. and *Olivancillaria carcellesi*. Finally, for the present there was greater diversity, with 31 species (16 bivalves and 15 gastropods), with the most abundant being the bivalves *Brachidontes rodriguezi*, *Glycymeris longior* and  *Amiantis purpurata*, and the gastropods *Crepidula* sp., *Tegula patagonica*, *Buccinanops globulosum*, *Bostrycapulus odites* and *Olivancillaria carcellesi*.

In this respect, it is believed that mollusk variations in the area are related partly to changes in temperature that have taken place since the Pleistocene (see Chap. 6), but are mostly associated with the presence of sub-environments of different energy levels within this bay.

Faunal changes in the cases described above are mainly related to local physical variations (i.e., substrate, availability of food and currents) connected to glacial periods such as the Last Glacial Maximum (ca. 24,000 years BP; Rabassa 2008). During these glacial periods, a large portion of the Atlantic Continental Shelf was

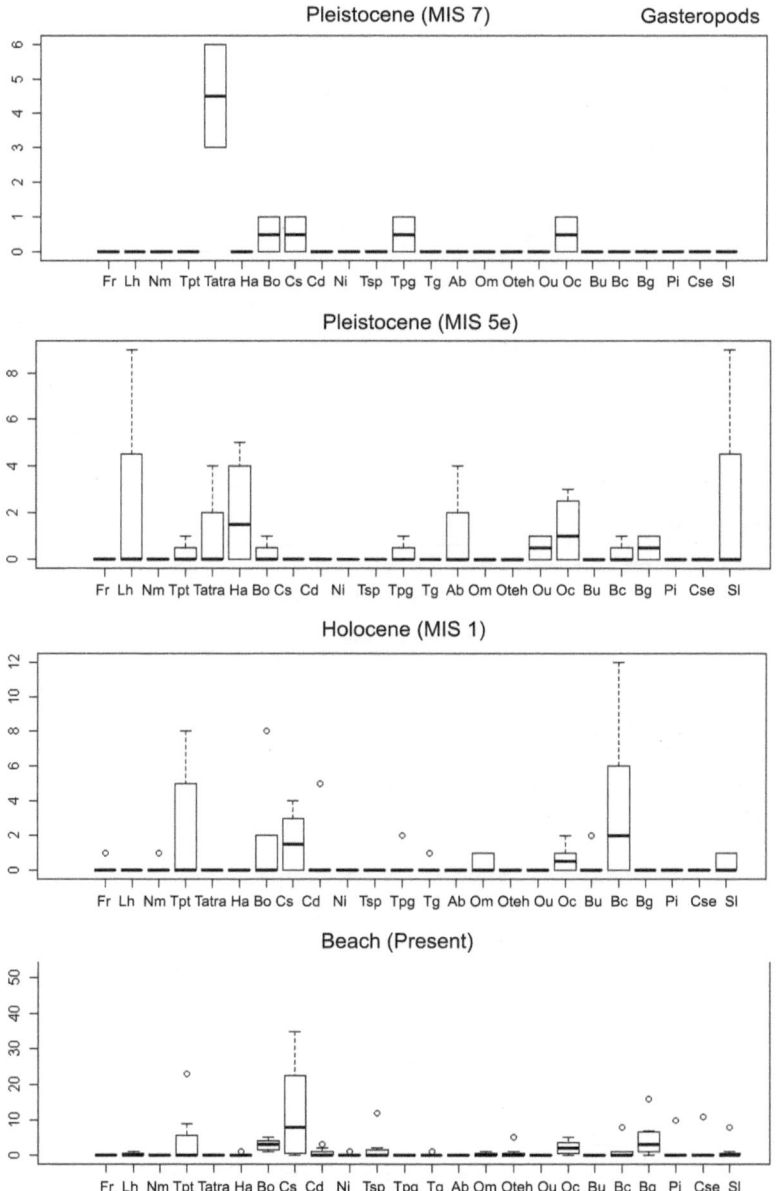

**Fig. 4.4**  Relative abundance of species of gastropods (**a**) and bivalves (**b**) from the northern San Matías Gulf at different time periods within the Quaternary

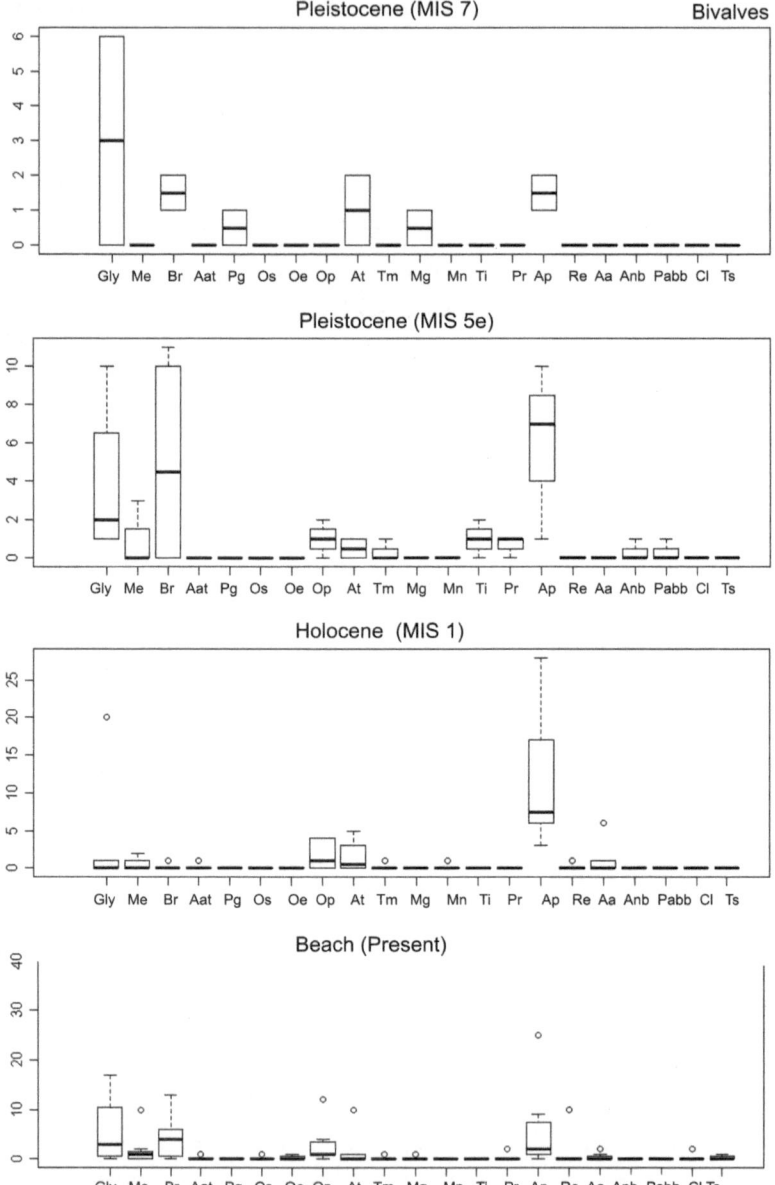

**Fig. 4.4** (continued)

exposed as a consequence of glacioeustatic movements, with the development of enormous plains along the Atlantic coast of Pampa and Patagonia (Rabassa et al. 2005; Ponce et al. 2011).

**Fig. 4.5** Faunistic changes
during the early-middle
Holocene in the Beagle
Channel. After deglaciation,
the first migrants arrived to
inhabit the vacant spaces,
with the subsequent
diversification and expansion
of the fauna which persists to
the present-day. **a** *Mulinia
edulis*, **b** *Mytilus* sp., **c** *Yoldia*
sp., **d** *Aulacomya atra*, **e** A
greater diversity of taxa

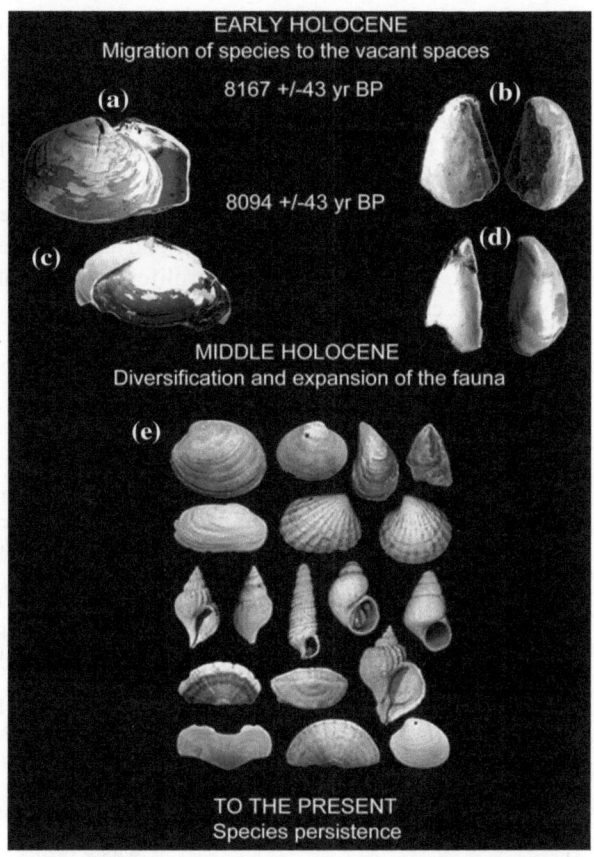

Once again in the Beagle Channel, but now within the early Holocene (ca. 8,000 years BP), new local communities developed in newly formed or recently vacated habitats through immigration of taxa from surrounding waters. After deglaciation, the first mollusks to arrive were eurytopic species such as *Mulinia edulis*, *Mytilus chilensis*, *Aulacomya atra* and *Yoldia* sp. (Rabassa et al. 2009; Gordillo et al. 2013), and then towards the middle Holocene the number of species increased (Fig. 4.5), and different local communities arose, depending on the physical characteristics that prevailed in each area. The first arriving taxa were species typical of tidal flats or areas more exposed to highly unstable conditions with longer episodes of exposure. In southern Chile, Velasco and Navarro (2003) demonstrated that *M. chilensis* and *M. edulis* exhibit a high degree of physiological plasticity. Reid and Osorio (2000) also mentioned a group of euryhaline taxa (including *M. chilensis*, *A. atra* and *M. edulis*) from a fjord system in southern Chile which tolerate sharp salinity gradients. Thus, organisms which represent the oldest marine stages of the Holocene tolerated large fluctuations in quality and quantity of suspended particulate matter due mainly to the resuspension of benthic sediment by the action of winds and tides. A phase of major expansion of fauna (ca. 7,500–5,000 years BP), along with further diversification of taxa (ca. 4,500–4,000 years BP), then indicates an evolution towards modern conditions. These changes in faunal composition from poor diversity to a more diverse fauna are related to local changes associated with the initial incoming of freshwater and the progressive input of marine waters (Gordillo et al. 2005). They represent an ecosystem transition which started with vacant niches first occupied by opportunistic species and/or eurithopic taxa, and their subsequent replacement by more diverse taxa, associated with the proliferation of habitats which occurred under truly marine conditions during the Holocene.

# References

Aitken AE (1990) Fossilization potential of Arctic fjord and continental shelf benthic macrofaunas. In: Dowdeswell JA, Scourse JD (eds) Glacimarine environments: processes and sediments. Geol Soc Lond Special Publ 53:155–176. doi:http://dx.doi.org/10.1144% 2FGSL.SP.1990.053.01.09

Archuby F, Adami, M, Martinelli J, Malvé M, Gordillo S (2011) Análisis de la fidelidad de las acumulaciones de valvas en playas de la Patagonia con respecto a las comunidades vivientes en el intermareal rocoso: implicancias tafonómicas. In: Resúmenes 8vo congreso latino-americano de malacología, Puerto Madryn

Boretto G, Gordillo S, Cioccale M, Colombo F, Fucks E (2013) Multi-proxy evidence of Late Quaternary environmental changes in the coastal area of Puerto Lobos (Northern Patagonia, Argentina). Quatern Int 305:188–205. doi:http://dx.doi.org/10.1016%2Fj.quaint.2013.02.017

Charó MP, Gordillo S, Fucks EE (2013) Paleoecological significance of Late Quaternary molluscan faunas of the Bahía San Blas area, Argentina. Quatern Int 301:135–149. doi:http:// dx.doi.org/10.1016%2Fj.quaint.2012.12.019

Feruglio E (1950) Descripción Geológica de La Patagonia. Dirección General de YPF, Buenos Aires

Gordillo S (1998) Distribución biogeográfica de los moluscos del Holoceno del litoral argentino-uruguayo. Ameghiniana 35:163–180

Gordillo S (1999) Holocene molluscan assemblages in the Magellan region. Scientia Marina 63(Suppl 1):15–22

Gordillo S (2009) Quaternary marine mollusks in Tierra del Fuego: insights from integrated taphonomic and paleoecologic analysis of shell assemblages in raised beaches. An Inst Pat 37:5–16

Gordillo S, Isla F (2011) Faunistic changes between the Middle/Late Pleistocene and the Holocene on the Atlantic coast of Tierra del Fuego: molluscan evidence. Quatern Int 233:101–112. doi:http://dx.doi.org/10.1016%2Fj.quaint.2010.06.006

Gordillo S, Coronato A, Rabassa J (2005) Quaternary molluscan faunas from the island of Tierra del Fuego after the last glacial maximum. Scientia Marina 69(Suppl 2):337–348

Gordillo S, Rabassa JO, Coronato A (2008) Paleoecology and paleobiogeographic patterns of Mid-Holocene mollusks from the Beagle channel (Southern Tierra del Fuego, Argentina). Rev Geol Chile 35:1–13. doi:http://dx.doi.org/10.4067/S0716-02082008000200007

Gordillo S, Cusminsky G, Bernasconi E, Ponce F, Rabassa J, Pino M (2010) Pleistocene marine calcareous macro-and-microfossils of Navarino Island (Chile) as environmental proxies during the last interglacial in southern South America. Quatern Int 221:159–174. doi:http://dx. doi.org/10.1016%2Fj.quaint.2009.10.025

Gordillo S, Bernasconi E, Cusminsky G, Coronato A, Rabassa J (2013) Late Quaternary environmental changes in southernmost South America reflected in marine calcareous macro- and-microfossils. Quatern Int 305:149–162. doi:http://dx.doi.org/10.1016%2Fj.quaint.2012. 11.016

Kidwell SM (2001) Preservation of species abundance in marine death assemblages. Science 294:1091–1094. doi:http://dx.doi.org/10.1126%2Fscience.1064539

Kidwell SM (2002) Time-averaged molluscan death assemblages: palimpsests of richness, snapshots of abundance. Geology 30:803–806. doi:http://dx.doi.org/10.1130%2F0091-7613% 282002%29030%3C0803%3ATAMDAP%3E2.0.CO%3B2

Lamy F, Rulemán C, Hebbeln D, Wefer G (2002) High-and low-latitude climate control on the position of the southern Perú-Chile current during the Holocene. Paleoceanography 17:1–10. doi:http://dx.doi.org/10.1029%2F2001PA000727

Ponce F, Rabassa J, Coronato A, Borromei A (2011) Paleogeographic evolution of the Atlantic coast of Pampa and Patagonia since the last glacial maximum to the middle Holocene. Biol J Linn Soc 103:363–379. doi:http://dx.doi.org/10.1111%2Fj.1095-8312.2011.01653.x

Rabassa J (2008) Late Cenozoic glaciations in Patagonia and Tierra del Fuego. In: Rabassa J (ed) The Late Cenozoic of Patagonia and Tierra del Fuego. Developments in Quaternary science, vol 11. Elsevier, Amsterdam, pp 151–204. doi:http://dx.doi.org/10.1016%2FS1571-0866% 2807%2910008-7

Rabassa J, Coronato A, Salemme M (2005) Chronology of the Late Cenozoic Patagonian glaciations and their correlation with biostratigraphic units of the Pampean region (Argentina). J S Am Earth Sci 20:81–103. doi:http://dx.doi.org/10.1016%2Fj.jsames.2005. 07.004

Rabassa J, Coronato A, Gordillo S, Candel M, Martinez M (2009) Paleoambientes litorales durante la trasgresión marina holocena en Bahía Lapataia, Canal Beagle, Parque Nacional Tierra del Fuego, Argentina. Rev Asoc Geol Argentina 65:648–659. doi:http://www.scielo. org.ar/pdf/raga/v65n4/v65n4a06.pdf

Reid DG, Osorio C (2000) The shallow-water marine mollusca of the Estero Elefantes and Laguna San Rafael, Southern Chile. Bull Natl Hist Mus Lond (Zool) 66:109–146

Root RB (1967) The niche exploitation pattern of the blue-gray gnatcatcher. Ecol Monogr 37:317–350. doi:http://dx.doi.org/10.2307%2F1942327

Velasco LA, Navarro JM (2003) Energetic balance of infaunal (*Mulinia edulis* King 1831) and epifaunal (*Mytilus chilensis* Hupé 1854) bivalves in response to wide variations in concentration and quality seston. J Exp Mar Biol Ecol 296:79–92. doi:http://dx.doi.org/10. 1016%2FS0022-0981%2803%2900316-2

# Chapter 5
# Biotic Interactions

**Abstract** Molluscan death assemblages are useful for reconstructing paleocommunities and can also provide signals for evaluating predation and other interactions. This chapter gives examples of biotic interactions that can be recorded in mollusk shells. We will provide data on the interaction between drilling predators and their shelled prey, and will also refer to preserved shell encrusters and organisms which settle on live hosts or dead shells.

**Keywords** Southern South America · Quaternary · Pleistocene · Holocene · Mollusca · Paleoecology · Drilling predation · Epibionts

Molluscan death assemblages are useful for reconstructing paleo communities and for evaluating faunistic changes. They also provide signals for evaluating predation and other biotic interactions, and therefore constitute an important tool for biological and paleontological studies.

## 5.1 Looking for Holes

Molluscan death assemblages provide direct evidence of biotic interactions, thus offering quantifiable data on predator-prey relationships (Kowalewski 2002). The fossil record yields abundant data on the interaction between drilling predators and their shelled prey, and is therefore an interesting venue for addressing evolutionary questions (Vermeij 1987; Kelley et al. 2003). Drilling predation by gastropods involves mechanical rasping with the radula, as well as secretions from the accessory boring organ (ABO) (Carriker 1981). Although it is known within several families of gastropods, most cases reported are produced by naticid and muricid gastropods.

In Argentinean Patagonia the most common muricid gastropod is *Trophon geversianus*, which inhabits both rocky and soft shallow bottoms. In southern Patagonia, this muricid gastropod preys upon mytilids and venerid clams (Gordillo

S. Gordillo et al., *Mollusk Shells as bio-geo-archives*,
South America and the Southern Hemisphere, DOI: 10.1007/978-3-319-03476-8_5,
© The Author(s) 2014

1998; Andrade and Ríos 2007; Gordillo and Archuby 2012a, b), depending on the predominant prey in the habitat in which it lives. *T. geversianus* always makes holes, but while it drills the valve walls of *M. chilensis*, it prefers to drill the valve edges of *A. atra* and *B. purpuratus*, with different characteristic patterns (Gordillo and Archuby 2012a). On *Tawera gayi* this whelk drills tronco–conical drillholes, perpendicular to the shell surface of its prey (Gordillo 1998; Gordillo and Amuchástegui 1998).

The case of the whelk *A. monodon,* which drills a hole in *B. purpuratus* but uses the outer lip of its shell as a wedge to open the valves of *M. chilensis* and *A. atra,* is also interesting.

These examples show that predatory damage to bivalve shells varies according to the predator and prey species, and also that techniques for attacking prey are highly specialized. In addition, preliminary results in Patagonia (Gordillo and Archuby 2012b, unpublished data) show an increase in the intensity of predation between the Pleistocene and present times. As pointed out by the authors, these changes should be analyzed taking into account the fact that Pleistocene shells were deposited prior to the formation of the San Matías and San Jorge gulfs (Ponce et al. 2011), thus resulting in variations of hydrological conditions and substrate types that probably affected the burial depth of these clams. The interactions also need to be analyzed along with other indicators of food levels in the past, such as growth rates and/or adult size, and used together with sedimentological information of Pleistocene deposits to help reconstruct relative patterns of paleoproductivity (Fig. 5.1).

## 5.2  Between Host and Guests

Continuing with biotic interactions within the marine realm, epibiosis is one of the few well-preserved biotic interactions in the fossil record dating back to the early Paleozoic (Lescinsky 2001). During the Cenozoic, shell encrusters and organisms able to settle on live and dead shells are very common (Taylor and Wilson 2003). Studying the spatial relationships between organisms that lived with each other provides insight into their community structure and environment. Epibiont composition is certainly useful for understanding patterns on live hosts, since these hosts act as 'discontinuous islands of substrate' within soft bottom environments. The study of epibiosis in death assemblages is also useful for understanding how a fossil assemblage differs from the living community from which it is derived, and for evaluating the correlation between living mollusks and recent dead and fossil shells.

Gordillo and Archuby (2012b) analyzed the presence of *Crepidula* spp. as an epibiont  of *Ameghinomya antiqua*. The epibiont was identified either by its presence or, if the shell was found separated from the host, it was recognized through a mark that reproduces the basal shape of the slipper snail. The analysis of epibiont distribution on *A. antiqua* shells, in conjunction with drill hole placement,

**Fig. 5.1** Examples of taxa, including bivalves and gastropods, showing holes produced by drilling gastropods. **a** *Ameghynomya antiqua*, **b** *Cyclocardia compressa*, **c** *Amiantis purpurata*, **d** *Calyptraea pileolus*, **e** *Pareuthria plumbea*, **f** *Tawera gayi*, **g** *Trophon geversianus*, **h** *Hiatella* sp., **i** *Nacella deaurata*. Scale: 1 cm

**Fig. 5.2** Diagram showing
the division of the outer shell
surface into seven arbitrary
regions (adapted from Ward
and Thorpe 1991)

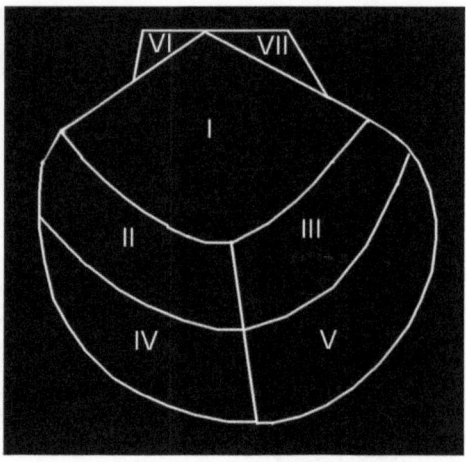

shows that they are more concentrated in the upper sector of the valve (75–71 %),
which might be explained by the vertical position and the semi-infaunal mode of
life of this clam. Nevertheless, the presence of drill holes in the lower sector of the
valve indicates that clams spent part of the time reclining on the sediment. There is
also evidence that clams with *Crepidula* spp. as commensals are less frequently
attacked by drilling gastropods.

Finally, another preliminary study (Gordillo et al. 2006) on modern and fossil
pectinid shells distinguished between 'live shells' and 'dead shells'. The term 'live
shell' (or live host) refers to shells which were encrusted or attached to while the
scallop was alive. It includes both living specimens and specimens that died
recently (i.e., dead specimens or empty, complete shells). The term 'dead shell' is
used to describe shells with post-mortem encrustation or attachment. On live
shells, the occurrence of the different major taxonomic groups was taken sepa-
rately for right (lower) and left (upper) valve surfaces, while on dead shells the
taxonomic composition of the epibionts was examined on the outer and inner valve
surfaces.

In addition, Gordillo et al. (2006) quantify taxa composition, abundance, spatial
distribution and the prevalence of epibionts associated with the scallop *Zygochlamys
patagonica* from southern Argentina, with the goal of analyzing changes in latitude
and time. For the analysis of spatial distribution each valve was arbitrarily divided
into seven areas (Fig. 5.2), roughly following the procedures of Ward and Thorpe
(1991), Sanfilippo (1994), and the percentage of coverage of each taxon or taxonomic group was estimated using a scale with three categories of surface coverage
for each area.

*Zygochlamys patagonica* shells from Patagonia provided a habitat for a great
variety of organisms (Fig. 5.3) that attach or encrustate as epibionts, but dead
shells exposed on the seafloor also acted as hard substrate for any surface-dwelling
organism.

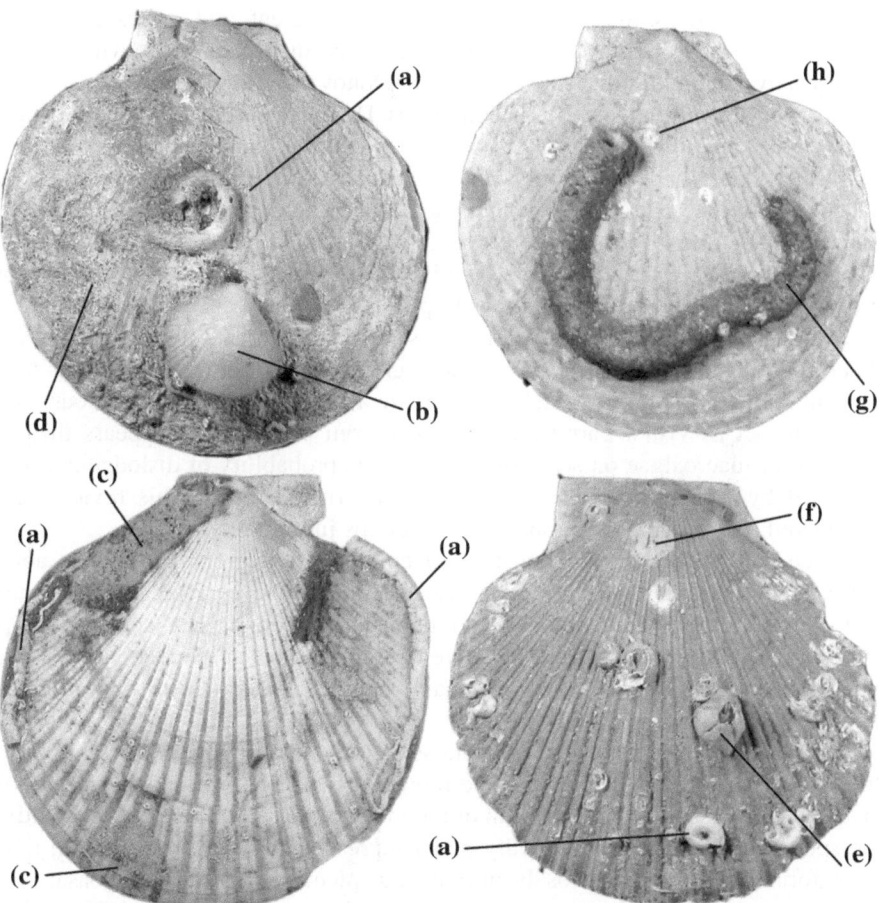

**Fig. 5.3** External view of recent *Zygochlamys patagonica* shells from Patagonia, Argentina, showing a variety of epibionts. **a** Calcareous serpulid, **b** Braquiopod, **c** Bryozoans, **d** Sponges, **e** Barnacle, **f** Scar left by a barnacle, **g** An agglutinated sabellid tube, **h** *Spirorbis* sp

Finally, it is important to note that when interactions between the host and their epibionts are considered, many different aspects (i.e., mortality, growth, predation and competition) must be taken into account.

For analyzing predation, in the case of scallops it should be remembered that free scallops normally swim to avoid predation. Sponges appear to have positive effects on their host since they camouflage the scallops and therefore protect them against predators (Forester 1979; Donovan et al. 2002). Furthermore, sponges increase survival by reducing the effects of sediment accumulation, and this has led several authors (Forester 1979; Pond 1992) to recognize a sponge-scallop mutualism. Farren (2003) demonstrated that sponges on scallops have chemical defenses that prevent both predation by sea stars and the settlement of barnacles. In the case of barnacles, the situation appears to be different, and encrustation could make scallops more susceptible to predators since they become heavier and are more easily captured since they soon tire (Donovan et al. 2003). By removing encrusting barnacles under aquaria conditions, Donovan et al. (2003) demonstrated that they strongly affected scallop swimming. Further information is needed to evaluate whether barnacles offer visual or tactile camouflage for predators. From data considered here it is possible that barnacles do not represent an obstacle since they are small, and they only covered a relatively small proportion of the shells. In any case, whether barnacles have positive or negative effects on scallops most probably depends on the weight a scallop is able to carry and still be capable of escaping by swimming. At least in mussels, field experiments by Buschbaum and Saier (2001) showed significantly lower growth with barnacle epibionts than without them. Other groups, such as epiphytic algae, hydroids and bryozoans also appear to act as visual barriers that protect from predators. It appears that the presence of macroalgae on scallops increases the probability of dislodgements, as observed by Gonzalez et al. (2001) after a storm. Based on this review, it is considered here that the epibionts affect scallops in a variety of ways.

When considering the relationships of epibionts it is probable that the abundance of different taxonomic groups is determined more by rates of recruitment and growth than by competition (McKinney and Jackson 1991; Ward and Thorpe 1991). In this sense, sponges are sensitive to high levels of suspended particles and many are excluded from areas with heavy sediment loads (Burns and Bingham 2002).

Taking into account the fact that most of the conspicuous taxonomic groups (i.e., bryozoans, and barnacles) can be found on fossil shells, the non-preservable taxa could represent a minor fraction of the total epibionts. One exception could be the sponges, which are undoubtedly under-represented in the fossil record; however, forms with fused or closely interlocked spicules or mineralized basal skeletons are common in the ancient (i.e., Paleozoic) fossil record (see Taylor and Wilson 2003). In our study on *Zygochlamys patagonica*, there were too few available fossil shells for any generalization on the absence of sponges in this assemblage, but other reasons, such as the preservation potential of the group in this kind of environment, might have played an important role. A final comment should be made on clionid sponges, which are absent from the live shells considered here, although they appear to be very common on dead shells. It is well known that clionid sponges occur worldwide, from polar seas to the tropics, and they have been reported on both living and dead shells (Wesche et al. 1997), including scallops (Del Río et al. 2001; Barthel et al. 1994). Clionid sponges, as shell borers, probably preferentially burrow through dead shells rather than live shells, and play an important role in the natural process of destruction of scallops, although this aspect requires further investigation. Forthcoming research will be focused on the taphonomic meaning of the epibionts on fossil Quaternary scallops from Southern Argentina.

# References

Andrade C, Ríos C (2007) Estudio experimental de los hábitos tróficos de *Trophon geversianus* (Pallas 1774) (Gastropoda: Muricidae): selección y manipulación de presas. An Inst Pat 35:45–53

Barthel D, Sundet J, Barthel K-G (1994) The boring sponge *Cliona vastifica* in a subarctic population of *Chlamys islandica*–an example of balanced commensalism. In: Van Soest R, Van Kempen T, Braekman JC (eds) Sponges in space and time. Balkema, Rotterdam, pp 289–296

Burns DO, Bingham, BL (2002) Epibiont sponges on the scallops *Chlamys hastata* and *Chlamys rubida*: increased survival in a high-sediment environment. J Mar Biol Ass UK 82:961–966. doi:http://dx.doi.org/10.1017%2FS0025315402006458

Buschbaum C, Saier B (2001) Growth of the mussel *Mytilus edulis* L. in the Wadden Sea affected by tidal emergence and barnacle epibionts. J Sea Res 45:27–36. doi:http://dx.doi.org/10.1016%2FS1385-1101%2800%2900061-7

Carriker MR (1981) Shell penetration and feeding by naticacean and muricacean predatory gastropods: a synthesis. Malacologia 20:403–422

Del Río C, Martinez SA, Scasso RA (2001) Nature and origin of spectacular marine miocene shell beds of northeastern Patagonia (Argentina): paleoecological and bathymetric significance. Palaios 16:3–25. doi:http://dx.doi.org/10.1669%2F0883-1351%282001%29016%3C0003%3ANAOOSM%3E2.0.CO%3B2

Donovan DA, Bingham BL, Farren HM, Gallardo R, Vigilant VL (2002) Effects of sponge encrustation on the swimming behaviour, energetics and morphometry of the scallop *Chlamys hastate*. J Mar Biol Ass UK 82:469–476. doi:http://dx.doi.org/10.1017%2FS0025315402005738

Donovan DA, Bingham BL, From M, Fleisch AF, Loomis ES (2003) Effects of barnacle encrustation on the swimming behaviour, energetics, morphometry, and drag coefficient of the scallop *Chlamys hastata*. J Mar Biol Ass UK 83:813–819. doi:http://dx.doi.org/10.1017%2FS0025315403007847h

Farren HM (2003) Sponge epibionts protect the scallop *Chlamys hastata* from barnacle encrustation and seastar predation, MS Dissertation Thesis, Western Washington University

Forester AJ (1979) The association between the sponge *Halichondria panacea* (Pallas) and scallop *Chlamys varia* (L.): a commensal-protective mutualism. J Exp Mar Biol Ecol 36:1–10. doi:http://dx.doi.org/10.1016%2F0022-0981%2879%2990096-0

Gonzalez SA, Stotz WB, Aguilar M (2001) Stranding of scallops related to epiphytic seaweeds on the coast of northern Chile. J Shellfish Res 20:85–88

Gordillo S, Bremec C, Mabragaña E, Schejter L (2006) Epibiontes de la vieira *Zygochlamys patagonica* (King y Broderip). Ocurrencia y Potencial de Preservación. In: Resúmenes del 9no congreso argentino de paleontología y bioestratigrafía. Academia Nacional de Ciencias, Córdoba

Gordillo S (1998) Trophonid gastropod predation on recent bivalves from the Magellanic Region. In: Johnston PA, Haggart JW (eds) Bivalves: an eon of evolution. Paleobiological Studies Honoring Norman N. Newell. University of Calgary Press, Calgary, pp 251–254

Gordillo S, Amuchástegui S (1998) Estrategias de depredación del gastrópodo perforador *Trophon geversianus* (Pallas) (Muricoidea: Trophonidae). Malacologia 39:83–91

Gordillo S, Archuby F (2012a) Predation by drilling gastropods and asteroids upon mussels in rocky shallow shores of southernmost South America: paleontological implications. Acta Palaeontol Pol 57:633–643. doi:http://dx.doi.org/10.4202%2Fapp.2010.0116

Gordillo S, Archuby S (2012b) Live-live and live-dead interactions in marine death assemblages: the case of the Patagonian clam *Venus antiqua*. Acta Palaeontol Pol. doi:http://dx.doi.org/10.4202/app.2011.0176

Kelley PH, Kowalewski M Hansen TA (2003) Predator-prey interactions in the fossil record. Kluwer Academic-Plenum Publishers, New York

Kowalewski M (2002) The fossil record of predation: an overview of analytical methods. Paleontol Soc Pap 8:3–42

Lescinsky HL (2001) Epibionts. In: Briggs DE, Crowther PR (eds) Palaeobiology II, Blackwell Science, Oxford, pp 460–464

McKinney FK, Jackson, JB (1991) Bryozoan evolution. University of Chicago Press, Chicago

Pond D (1992) Protective-commensal mutualism between the queen scallop *Chlamys opercularis* (Linnaeus) and the encrusting sponge *Suberites*: J Mollusc Stud 58:127–134. doi:http://dx.doi.org/10.1093%2Fmollus%2F58.2.127

Ponce JF, Rabassa J, Coronato A, Borromei AM (2011) Palaeogeographical evolution of the Atlantic coast of Pampa and Patagonia from the last glacial maximum to the Middle Holocene. Biol J Linn Soc 103:363–379. doi:http://dx.doi.org/10.1111%2Fj.1095-8312.2011.01653.x

Sanfilippo R (1994) Polychaete distribution patterns on *Chlamys patagonica* of the Magellan Straits. Mém Mus Nat D'Hist Nat 162:535–540

Taylor PD, Wilson MA (2003) Palaeoecology and evolution of marine hard substrate communities. Earth-Sci Rev 62:1–103. doi:http://dx.doi.org/10.1016%2FS0012-8252%2802%2900131-9

Vermeij GJ (1987) Evolution and escalation: an ecological history of life. Princeton University Press, Princeton

Ward MA, Thorpe JP (1991) Distribution of encrusting bryozoans and other epifauna on the subtidal bivalve *Chlamys opercularis*. Mar Biol 110:253–259

Wesche SJ, Adland RD, Hooper JN (1997) The first incidence of clionid sponges (Porifera) from the Sydney rock oyster *Saccostrea commercialis* (Iredale and Roughley 1933). Aquaculture 157:173–180. doi:http://dx.doi.org/10.1016%2FS0044-8486%2897%2900139-7

# Chapter 6
# Biogeography

**Abstract** The study area in southern South America comprises a very extensive geographic area covering more than 2,000 km and involving two distinct biogeographic areas. This chapter focuses on faunistic changes within and between provinces, faunistic shifts and the extinction of particular species during the Quaternary.

**Keywords** Southern South America · Quaternary · Pleistocene · Holocene · Mollusca · Faunal distribution · Faunal shifts · Extinction · Transoceanic migrants

Our research is focused on the Quaternary mollusks of a very extensive geographic area covering more than 2,000 km. It involves two distinct biogeographic areas: the Magellan (43–55° S) and the Argentinean (30–43° S) provinces. The boundary between these provinces has not been clearly established since it varies seasonally between 41 and 43° S, with northward fluctuations of the cold Malvinas Current in winter (Balech and Ehrlich 2008). The dominance of different water masses in these two provinces causes water temperatures to be lower in the Magellan Province (3.5–11 °C, Boltovskoy 1979) than in the Argentinean Province (18–24 °C, Boltovskoy 1979). As marine faunal distribution is mainly related to global water temperature gradients (Valentine et al. 1978; Valentine and Jablonski 1985), these water temperature differences are one of the main determinants of the molluscan fauna composition of these two regions (Gordillo 1998; Pastorino 2000) (Fig. 6.1).

## 6.1 Faunistic Changes Within and Between Provinces

It was observed that the changes in the quantitative and qualitative analyses between Pleistocene and Holocene mollusks from the study region are mostly due to the presence of sub-environments and local, rather than global, environmental changes that affect ecological parameters.

S. Gordillo et al., *Mollusk Shells as bio-geo-archives*,
South America and the Southern Hemisphere, DOI: 10.1007/978-3-319-03476-8_6,
© The Author(s) 2014

**Fig. 6.1** Magellan, Argentinean and Brazilian malacological provinces in South America

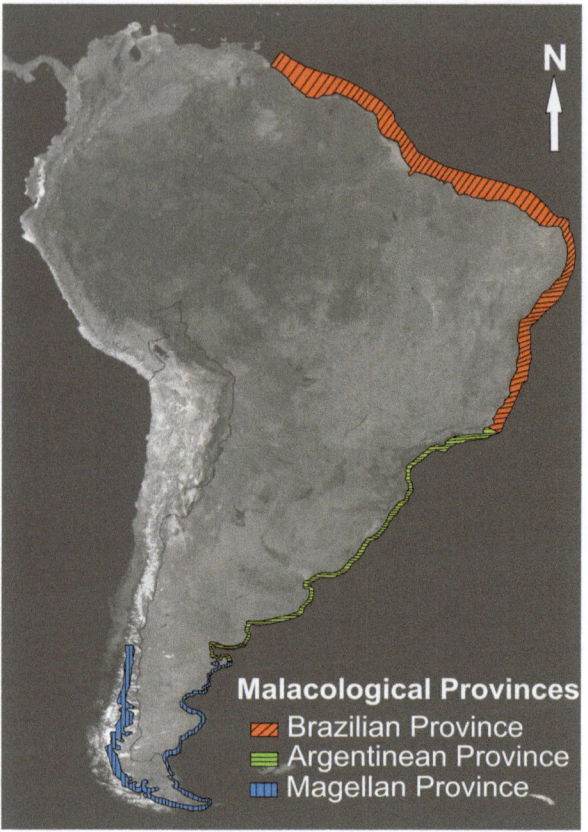

For example, in the region of the Colorado River Delta (39° S, south of Buenos Aires Province, Argentina) changes were observed in the proportion of warm temperate species between the MIS9 and the present (Fig. 6.2). There has been an increase in the proportion of warm water gastropods from MIS9 to the present, whereas the proportion of warm water bivalves has slightly decreased in the present.

These changes are partially linked to shifts in the boundaries of the areal extent of some organisms, and partly to local environmental changes.

A notable example of a species that changed its range of distribution during the Quaternary is the infaunal bivalve *Anomalocardia brasiliana*. This was mentioned by Charó et al. (2013c) in the San Matías Gulf for the MIS5e (Fig. 6.3), but is today distributed further north from the French Antilles (18° N) to the coast of Brazil (33° S), thus indicating warmer conditions at that time, coinciding with global changes. Other species of warm lineage that also expanded their range of distribution during the Pleistocene were *Crassostrea rizophorae* and *Abra aequalis* (Charó et al. 2013a, b).

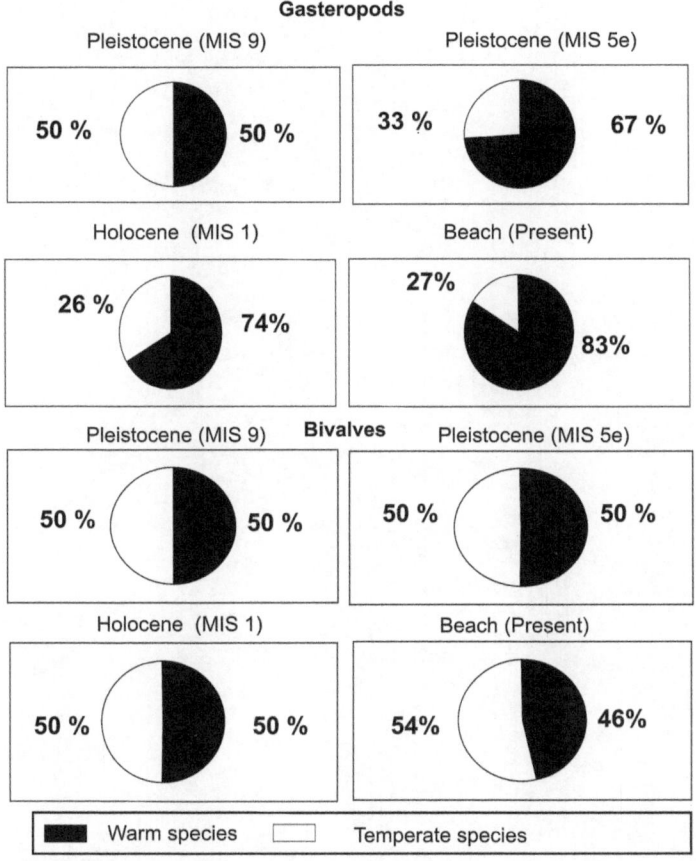

**Fig. 6.2** Proportion of warm versus temperate taxa in southern Buenos Aires Province, at different time periods during the Quaternary

Faunal analysis was carried out by Boretto et al. (2013) in the Pleistocene-Holocene beach ridge system of Puerto Lobos (42°00′S, 65°4′W; northern Chubut Province, Argentina), which is a site located within the border area between the two biogeographic provinces. The analysis showed that the faunal turnover record could be related to a migration of the boundary between the Argentinean and Magellan provinces. The Argentinean Province borders the Brazilian Province to the north (linked to the Brazil Current) and the Magellan Province to the south (linked to the Malvinas Current in the Atlantic Ocean and the Humboldt Current in the Pacific) (Boschi 2000; Martínez et al. 2011). Thus, the existence of thermal gradients in the SW Atlantic produced by the arrival of the warm Brazil Current from the north and the cold Malvinas Current from the south, means that their boundaries are not static due to the influence exerted by water masses of different depths at the same latitude, and to variations caused by ocean–atmosphere

**Fig. 6.3** Map showing the location of a Pleistocene fossil record of *Anomalocardia brasiliana*, further south than its present distribution along the Brazilian Province to the north

interaction (Martínez et al. 2011). Recently, Muhs et al. (2012) showed how climate changes during the Quaternary have affected the distribution range of mollusks from San Nicolas Island, on the boundary between the Californian and Oregonian marine invertebrate faunal provinces, which are dominated by the cold southward California Current and the warm Inshore Countercurrent, respectively.

The youngest Puerto Lobos Holocene deposit is from ca. 1564 AD, which was deposited before the Maunder Minimum of the Little Ice Age (LIA). There was a period of prolonged solar quiet from about 1645 until 1715, and the coincidence of a prolonged "solar minimum" with the coldest excursion of the LIA, which has been noted by many who have looked at the possible relationship between the sun and terrestrial climate (Eddy 1976). Evidence of this event has been registered in continental areas of Argentina (Cioccale 1999; Glasser et al. 2002; Rabassa 2008; Strelin et al. 2008; Masiokas et al. 2009).

Boretto et al. (2013) showed some of the differences between the Pleistocene and the Holocene. The most relevant taxonomic differences are the presence of *T. atra* and *M. patagonica* in Pleistocene ridges and the diversification of taxa in the Holocene. The coastal area of Puerto Lobos has also recorded a faunal shift which occurred during the late Holocene. The data shows that at present there is a greater proportion of taxa typical of the cold-water Magellan Province, while during the Holocene the most typical element in the area was *G. longior*, characteristic of northern Argentina (Argentinean Province). This difference can be explained by a shift in the environmental conditions after 1564 AD, which is the age of the youngest beach ridge, when the southern limit of the Argentinean Province retracted. This has been calculated using the Bayarsky and Codignotto (1982) radiocarbon age, and calibrated considering the reservoir effect, as shown by Dubois (2009).

Because of this, species belonging to the Magellan Province displace the fauna of the Argentinean Province to the north, where the sea surface temperature reflects a greater influence of the warm Brazilian Current.

## 6.2 Some of Them Have Gone

In this section we refer to certain taxa that appear in Pleistocene sediments but then became extinct (Fig. 6.4). These are the bivalves *Chama iudicai* (Pastorino 1991) and *Glycymeris sanmatiensis* (Bayer and Gordillo 2013) from northern Patagonia, and the muricid gastropod *Lepsiella ukika* (Gordillo and Nielsen 2013) from the southernmost tip of South America.

A special case is the gastropod *Tegula atra,* which appears as a living species in the Pacific, but only as a Pleistocene fossil in the Atlantic. As a living species, it exhibits a wide range of distribution from Peru to Chile; but on the Atlantic side, it only appears in Pleistocene marine deposits, and then became extinct perhaps before the Holocene. In northern Patagonia, this species was found in Pleistocene sediments associated with other temperate to warm-water species like *Amiantis purpurata* and *Glycymeris longior*, along with two extinct *Chama iudicai* and *Glycymeris sanmatiensis*. However, Aguirre et al. (2013) characterize this gastropod as a cold water species and attribute its disappearance in the Atlantic to oceanographic changes related to the influence of the Brazil Current or to the lower intensity of the Malvinas Current after the Last Glacial Maximum. At present there is still no convincing explanation of how oceanographic changes that occurred in the late Pleistocene may have affected the larval stage (settlement) and/or the adult stage of this species that lives mainly on macroalgae, on which it feeds. Only on the basis of interdisciplinary studies, including environmental variables, genetics and the ecophysiological constraints of living species, as well as the fossil record of the genus *Tegula* in South America, a conclusion can be

**Fig. 6.4** External view of the Quaternary extinct taxa in southern South America. **a, b** *Chama iudicai*; **c, d** *Tegula atra*; **e, f** *Glycymeris sanmatiensis*; **g, h** *Lepsiella ukika*. *Scale bar* 1 cm

reached about the relationship between *Tegula atra* and its appearance in Pleistocene marine deposits along the Atlantic in Patagonia, Argentina. It is also possible that there has been a misunderstanding between researchers, and that the Pleistocene extinct Atlantic species is not in fact the same as the living Pacific species. If so, this would be a case of morphological convergence of the shell, which has been noted by De Francesco (2007) as one of the limitations in determining Quaternary species from the fossil record.

In either case, *Tegula atra* in conjunction with the two northern Patagonian species (*Chama iudicai* and *Glycymeris sanmatiensis*) lived during Pleistocene times before the formation of the San Matías Gulf (Ponce et al. 2011; Isla 2013). This gulf is a semicircular basin separated from the open sea by areas that have developed a threshold of lesser depth than the central part of the basin (Cavallotto 2008). The origin of this gulf, and others from Patagonia, may be related to ancient continental depressions of hydro-aeolian origin which were later flooded by the sea, probably with the formation of large interior lakes, rather than having tectonic origins (Mouzo et al. 1978; Cavallotto and Violante 2003). On the basis of the analysis of this palaeogeographical evolutionary model, Ponce et al. (2011) concluded that the San Matías and San José gulfs would have been formed at around 12,000 calibrated years BP. A radiocarbon dating performed on Late Pleistocene wood fragments extracted from the bottom confirmed an age of between 11,500 and 11,000 years, when the sea level surpassed the sill of the gulf (today 50 m below mean sea level) during postglacial sea-level rise (Isla 2013).

Based on these studies, and together with the fact that prior to the Last Glacial Maximum (24 kya BP; Rabassa 2008), the sea level was 120 m below the present level, we have deduced that during the deposition time of these Pleistocene bivalve species, this area was a huge coastal plain, more exposed to the open sea, with a different hydrological regime and subject to storm effects and rapid sedimentation rates. In this respect, such geomorphological and sedimentological changes are likely to have influenced ecological patterns of benthic marine communities.

Moreover, the presence of *Ch. iudicai* and *G. sanmatiensis* in late Pleistocene (MIS5e) marine deposits from the San Matías Gulf, together with changes in the southern extension of certain species of warm lineage (i.e., *Anomalocardia brasiliana*, *Crassostrea rizophorae* and *Abra aequalis*) strongly suggests at least slightly warmer sea temperatures for the late Pleistocene in comparison with the Holocene.

Based on previous studies in other regions of the world and along the coast of Argentina, we can estimate that the last interglacial (MIS5e) would have been characterized by an increase in the global mean surface temperature of at least 2 °C, thus making it warmer than at present (Murray-Wallace and Belperio 1991; Murray-Wallace et al. 2000; Rohling et al. 2008), and also that the sea level was approximately 5–6 m higher than the present sea level (Shackleton 1987; Neumann and Hearty 1996).

This event would have been associated with changes in the distribution of species, with warmer marine records during the Pleistocene in different parts of the world (e.g., Cuerda et al. 1991; Muhs et al. 2002; Zazo et al. 2010, among others), and with other changes in South America, as mentioned by Chaar and Farinati (1988), and Rojas and Urteaga (2011).

The last extinct species is *Lepsiella ukika* (Gordillo and Nielsen 2013), which was recovered from a late Pleistocene marine raised beach located on Navarino Island, in southern South America. This is the first finding of this genus in the Americas, since it was previously geographically confined to New Zealand and to the temperate coast of Australia.

As mentioned in Sect. 4.3, during the Quaternary period, the southern tip of South America was affected by several glaciations which might have excluded much of the benthic marine fauna inhabiting this region, with the consequent interruption (more than once) of the connection between the Atlantic and the Pacific Oceans (see Gordillo 2009). These glaciations also shaped the receptive southern South American fjord region, which is considered a major reason for the high biodiversity of the region (Kiel and Nielsen 2010).

Within this in mind, the Navarino records of *Lepsiella* probably belong to a stock (or a short lived pioneer) derived from an Australasian population. It is plausible that the arrival of this taxon into southern South America took place during a glacial period prior to MIS4 to MIS2. The polar front might have shifted northwards during glacial times (Fraser et al. 2009), and the shallower, more northern position of the Antarctic Circumpolar Current (ACC) would have facilitated the circumpolar journey of this taxon from Australasia to South America.

Later, during the last interglacial (MIS5) ca. 125,000 years ago, this species remained in the Beagle Channel. At that time, a rich fauna including foraminifers, ostracods and mollusks developed in this channel (see Gordillo et al. 2010). After that, during the last glaciation, marine taxa living in this interior channel were separated and survived in marine refuges. Temperate taxa such as *Lepsiella* therefore disappeared during this period. When climatic conditions improved, most taxa reoccupied the ecological niches from marine refuges. The extinction of *Lepsiella* in the Beagle Channel could perhaps be related to the fact that this pioneer or derived population of *Lepsiella* was eliminated by competition. This assumption (see Gordillo and Nielsen 2013) is based on the fact that *Lepsiella* was found with numerous other predatory muricids, which implies that these species were contemporaneous and therefore had to compete for food. Another possibility, more difficult to prove, would be the lack of a suitable habitat, although it is quite plausible that as a result of sea-level fluctuations rocky shores at the intertidal level temporarily disappeared.

## 6.3  Transoceanic Migrants During the Quaternary

Beu et al. (1997) showed that the migration of mollusks via the ACC from South America to New Zealand might have occurred during the most extreme Pleistocene glaciations when this current was displaced northward. Additionally, a large burst of dispersed genus-group taxa arrived in South America during the Late

Oligocene–Early Miocene period, with virtually no further dispersal from New Zealand to South America. However, other mollusks could have followed a reverse route from Australasia to South America, as in the case of the venerid bivalve *Tawera*, which presumably reached South America during the Quaternary (Gordillo 2006).

Another example of Pleistocene transoceanic incursions of mollusks, in this case from South America to South Africa, is the muricid *Concholepas concholepas,* which only lives in South America but has been recorded in Late Pleistocene coastal deposits in southern South West Africa-Namibia (Kensley 1985).

Finally, the finding of *Lepsiella ukika* in southern South America is explained on the basis of transoceanic migration from Australasia by means of the ACC, perhaps during a Quaternary glacial period (see Gordillo and Nielsen 2013). Based on three different arguments (shell morphology, water masses and oceanic circulation, and ecology) these authors highlight the fact that this non-planktonic muricid was able to migrate, potentially using kelp as raft, to the southern tip of South America. This alternative dispersal mechanism of non-planktonic taxa is also potentially applicable to other fossils with a disjunct distribution.

Recent genetic studies on marine biota reinforce these interpretations of the distribution of fossil faunas, since they show increasing evidence that populations of organisms without planktonic larval stages can also be widely dispersed, and that rafting is most frequently involved (Thiel and Gutow 2005a, b; Thiel and Haye 2006; Macaya 2010; Nikula et al. 2010). For example, using the DNA barcoding method, Macaya (2010) provides strong evidence that gene flow along the Southern Ocean is occurring over ecological time scales, where rafting of detached reproductive *Macrocystis* kelp seems to be facilitated by the ACC connecting populations in the Southern Hemisphere. This author also gives evidence suggesting that kelp rafts act as an important dispersal mechanism for this species, thus providing vital information on the factors which are shaping the evolution of the largest seaweed on Earth. Concerning the associated fauna, Nikula et al. (2010) demonstrated that long-distance oceanic rafting explains the broad geographic distribution of two crustaceans across the subantarctic. Additionally, in a recent molecular study, Fraser et al. (2009) pointed out that the giant seaweed *Durvillea* has an extremely high dispersal potential since it is capable of rafting for vast distances. They concluded that *Durvillea* in southern Chile originated from source populations in New Zealand. Moreover, due to the local eddies, estuarine fronts and internal waves, the channels and fjords of southern South America appear to act as an extensive retention zone for floating items such as rafted kelp, which accumulated in the internal waters (Hinojosa et al. 2010, 2011). These authors pointed out that retention zones near the oceanic end of the channels may trap *Durvillea* rafts coming from coastal or distal oceanic sources. Based on these considerations, Gordillo and Nielsen (2013) suggest that kelp rafting constitutes a means of transportation for muricid gastropods, thus giving them a better chance of extending their range or of migrating from one region to another. These gastropods could therefore travel huge distances among kelp holdfasts of *Macrocystis*, *Durvillea*, or other macroalgae.

As can be seen, the importance of addressing interdisciplinary studies when considering Quaternary fauna is clear.

# References

Aguirre ML, Richiano S, Donato M, Farinati EA (2013) *Tegula atra* (Lesson, 1830) (Mollusca, Gastropoda) in the marine Quaternary of Patagonia (Argentina, SW Atlantic): Biostratigraphical tool and palaeoclimate-palaeoceanographical signal. Quatern Int 305:163–187. doi:http://dx.doi.org/10.1016/j.quaint.2013.02.011

Balech E, Ehrlich M (2008) Esquema biogeográfico del mar Argentino. Rev Inv Des Pesqu 19:45–75

Bayarsky A, Codignotto JO (1982) Pleistoceno Holoceno marino en Puerto Lobos, Chubut. Rev Asoc Geol Arg 37:91–99

Bayer S, Gordillo S (2013) A new Pleistocene species of *Glycymeris* (Bivalvia, Glycymerididae) from northern Patagonia, Argentina. Ameghiniana 50:265–268. doi:http://dx.doi.org/10.5710/AMGH.28.02.2013.578

Beu AG, Griffin M, Maxwell PA (1997) Opening of Drake Passage gateway and Late Miocene to Pleistocene cooling reflected in Southern Ocean molluscan dispersal: evidence from New Zealand and Argentina. Tectonophysics 281:83–97. doi:http://dx.doi.org/10.1016%2FS0040-1951%2897%2900160-1

Boltovskoy E (1979) Palaeoceanografía del Atlántico Sudoccidental desde el Mioceno según estudios foraminiferológicos. Ameghiniana 16:357–389

Boschi EE (2000) Species of decapod crustaceans and their distribution in the American marine zoogeographic provinces. Rev Inv Des Pesque 13:1–136

Boretto G, Gordillo S, Cioccale M, Colombo F, Fucks E (2013) Multi-proxy evidence of Late Quaternary environmental changes in the coastal area of Puerto Lobos (Northern Patagonia, Argentina). Quatern Int 305:188–205. doi:http://dx.doi.org/10.1016%2Fj.quaint.2013.02.017

Cavallotto JL (2008) Geología y geomorfología de los ambientes costeros y marinos. In: Boltovskoy D (ed) Atlas de Sensibilidad Ambiental de la Costa y el Mar Argentino. DVD Secretaría de Ambiente y Desarrollo Sustentable de la Nación. Buenos Aires. http://atlas.ambiente.gov.ar/

Cavallotto JL, Violante RA (2003) Late Pleistocene and Holocene transgressions in the northern Patagonian gulfs, Argentina. Continental shelves during the last glacial cycle, IGCP 464, Gdansk, pp 69–70

Charó MP, Fucks EE, Gordillo S (2013a) Moluscos bentónicos marinos del Cuaternario de Bahía Anegada (Sur de Buenos Aires, Argentina): variaciones faunísticas en el Pleistoceno tardío y Holoceno. Rev Mex Cs Geol 30:404–416

Charó MP, Gordillo S, Fucks EE (2013b) Paleoecological significance of Late Quaternary molluscan faunas of the Bahía San Blas area, Argentina. Quatern Int 301:135–149. doi:http://dx.doi.org/10.1016%2Fj.quaint.2012.12.019

Charó MP, Gordillo S, Fucks EE (2013c) Primer hallazgo de *Anomalocardia brasiliana* (Gmelin 1791) en el interglacial MIS5E en el sitio Baliza Camino (Bahía San Antonio, Río Negro, Argentina). In: Resúmenes del 1er congreso argentino de malacología, La Plata

Chaar E, Farinati E (1988) Evidencias paleontológicas y sedimentológicas de un nivel marino pleistoceno en Bahía Blanca, provincia de Buenos Aires, Argentina. In: Actas de la 2da jornada geológica bonaerense, Bahía Blanca

Cioccale MA (1999) Climatic fluctuations in the central region of Argentina in the last 1000 years. Quatern Int 62:35–47. doi:http://dx.doi.org/10.1016%2FS1040-6182%2899%2900021-X

Cuerda J, Vicens D, Gracia F (1991) Malacofauna y estratigrafía del Pleistoceno Superior marino de San Real (Santa Margalida, Mallorca). Boll Soc Hist Nat Balears 34:99–108

De Francesco C (2007) Las limitaciones a la identificación de especies de *Heleobia* Stimpson, 1865 (Gastropoda: Rissooidea) en el registro fósil del Cuaternario tardío y sus implicancias paleoambientales. Ameghiniana 44:631–635

Dubois C (2009) Valores de efecto reservorio marino para los últimos 5.000 años obtenidos en concheros de la costa Atlántica Norpatagónica (Golfo San Matías, Argentina). Magallania 37:139–147

Eddy JA (1976) The Maunder minimum. Science 192:1189–1202. doi:http://dx.doi.org/10.1126%2Fscience.192.4245.1189

Fraser CI, Nikula, R, Spencer, HG, Waters, JM (2009) Kelp genes reveal effects of subantarctic sea ice during the Last Glacial Maximum. Proc Natl Acad Sci U S A 106:3249–3253. doi:http://dx.doi.org/10.1073%2Fpnas.0810635106

Glasser NF, Hambrey MJ, Aniya M (2002) An advance of soler glacier, north Patagonian Icefield. Holocene 12:113–120. doi:http://dx.doi.org/10.1191%2F0959683602hl526rr

Gordillo S (1998) Distribución biogeográfica de los moluscos del Holoceno del litoral argentino-uruguayo. Ameghiniana 35:163–180

Gordillo S (2006) The presence of *Tawera gayi* (Hupé in Gay, 1854) (Veneridae, Bivalvia) in southern South America: Did *Tawera* achieve a Late Cenozoic circumpolar traverse? Palaeogeogr Palaeoc 240:587–601. doi:http://dx.doi.org/10.1016%2Fj.palaeo.2006.03.009

Gordillo S (2009) Quaternary marine mollusks in Tierra del Fuego: insights from integrated taphonomic and paleoecologic analysis of shell assemblages in raised beaches. An Inst Pat 37:5–16

Gordillo S, Nielsen S (2013) The Australasian muricid gastropod *Lepsiella* as Pleistocene visitor to southernmost South America. Acta Palaeontol Pol 58:777–783. doi:http://dx.doi.org/10.4202%2Fapp.2011.0186

Gordillo S, Cusminsky G, Bernasconi E, Ponce JF, Rabassa JO, Pino M (2010) Pleistocene marine calcareous macro-and-microfossils of Navarino Island (Chile) as environmental proxies during the last interglacial in southern South America. Quatern Int 221:159–174. doi:http://dx.doi.org/10.1016%2Fj.quaint.2009.10.025

Hinojosa IA, Pizarro M, Ramos M, Thiel M (2010) Spatial and temporal distribution of floating kelp in the channels and fjords of southern Chile. Estuar Coast Shelf S 87:367–377. doi:http://dx.doi.org/10.1016%2Fj.ecss.2009.12.010

Hinojosa IA, Rivadeneira MM, Thiel M (2011) Temporal and spatial distribution of floating objects in coastal waters of central-southern Chile and Patagonia. Cont Shelf Res 31:172–186

Isla F (2013) The flooding of the San Matías Gulf: the Northern Patagonia sea-level curve. Geomorphology (in press). doi:http://dx.doi.org/10.1016%2Fj.geomorph.2013.02.013

Kensley B (1985) The fossil occurrence in southern Africa of the South American intertidal mollusk *Concholepas concholepas*. An S Afr Mus 97:1–7

Kiel S, Nielsen SN (2010) Quaternary origin of the inverse latitudinal diversity gradient among southern Chilean mollusks. Geology 38:955–958. doi:http://dx.doi.org/10.1130%2FG31282.1

Macaya E (2010) Phylogeny, connectivity and dispersal patterns of the giant kelp *Macrocystis* (Phaeophyceae), PhD dissertation thesis, Victoria University of Wellington, Wellington

Martínez S, Del Río CJ, Rojas A (2011) Bases paleontológicas de la biodiversidad de moluscos del Atlántico Sudoeste. In: Souza I (ed) Paleontologia: Cenários da Vida. Editorial Interciência, pp 107–117

Masiokas MH, Luckman BH, Villalba R, Delgado S, Skvarca P, Ripalta A (2009) Little Ice Age fluctuations of small glaciers in the Monte Fitz Roy and Lago del Desierto areas, south Patagonian Andes, Argentina. Palaeogeogr Palaeocl 281:351–362. doi:http://dx.doi.org/10.1016%2Fj.palaeo.2007.10.031

Mouzo F, Garza ML, Izquierdo JF, Zibecchi RO (1978) Rasgos de la Geología del Golfo Nuevo (Chubut). Acta Oceanogr Arg 2:69–91

Muhs DR, Simmons KR, Kennedy GL, Rockwell TR (2002) The last interglacial period on the Pacific coast of North America. Geol Soc Am Bull 114:569–592. doi:http://dx.doi.org/10.1130%2F0016-7606%282002%29114%3C0569%3ATLIPOT%3E2.0.CO%3B2

Muhs DR, Simmons KR, Schumann R, Groves LT, Mitrovica JX, Laurel D (2012) Sea-level history during the Last Interglacial complex on San Nicolas Island, California: implications for glacial isostatic adjustment processes, paleozoogeography and tectonics. Quatern Sci Rev 37:1–25

Murray-Wallace CV, Belperio AP (1991) The last interglacial shoreline in Australia—a review. Quat Sci Rev 10:441–446. doi:http://dx.doi.org/10.1016%2F0277-3791%2891%2990006-G

Murray-Wallace CV, Beu AG, Kendric GW, Brown LJ, Belperio AP, Sherwood JE (2000) Palaeoclimatic implication of the occurrence of the arcoid bivalve *Anadara trapezia* (Deshayes) in the Quaternary of Australasia. Quat Sci Rev 19:559–590. doi:http://dx.doi.org/10.1016%2FS0277-3791%2899%2900015-3

Neumann AC, Hearty PJ (1996) Rapid sea-level changes at the close of the last interglacial (substage 5e) recorder in Bahamian island. Geology 24:775–778

Nikula R, Fraser CI, Spencer HG, Waters JM (2010) Circumpolar dispersal by rafting in two subantarctic kelp-dwelling crustaceans. Mar Ecol-Prog Ser 405:221–230. doi:http://dx.doi.org/10.3354%2Fmeps08523

Pastorino G (1991) The genus *Chama* Linné (Bivalvia) in the Marine Quaternary of northern Patagonia, Argentina. J Paleont 65:756–760

Pastorino G (2000) Asociaciones de moluscos de las terrazas marinas cuaternarias de Río Negro, Argentina. Ameghiniana 37:131–156

Ponce JF, Rabassa J, Coronato A, Borromei AM (2011) Palaeogeographical evolution of the Atlantic coast of Pampa and Patagonia from the last glacial maximum to the Middle Holocene. Biol J Linn Soc 103:363–379. doi:http://dx.doi.org/10.1111%2Fj.1095-8312.2011.01653.x

Rabassa J (2008) Late Cenozoic glaciations in Patagonia and Tierra del Fuego. In: Rabassa J (ed) The Late Cenozoic of Patagonia and Tierra del Fuego. Developments in Quaternary Science, vol 11. Elsevier, Amsterdam, pp 151–204. doi:http://dx.doi.org/10.1016%2FS1571-0866%2807%2910008-7

Rohling EJ, Grant K, Hemleben CH, Siddall M, Hoogakker BAA, Bolshow M, Kucera M (2008) High rates of sea-level rise during the last interglacial period. Nat Geosci 1:38–42. doi:http://dx.doi.org/10.1038%2Fngeo.2007.28

Rojas A, Urteaga D (2011) Late Pleistocene and Holocene chitons (Mollusca, Polyplacophora) from Uruguay: palaeobiogeography and palaeoenvironmental reconstruction in mid latitudes of the southwestern Atlantic. Geobios 44:377–386. doi:http://dx.doi.org/10.1016%2Fj.geobios.2010.09.002

Shackleton NJ (1987) Oxygen isotopes, ice volumes and sea level. Quat Sci Rev 6:183–190. doi:http://dx.doi.org/10.1016%2F0277-3791%2887%2990003-5

Strelin J, Casassa G, Rosqvist G, Holmlund P (2008) Holocene glaciations in the Ema Glacier Valley, Monte Sarmiento Massif, Tierra del Fuego. Palaeogeogr Palaeocl 260: 299–314. doi:http://dx.doi.org/10.1016%2Fj.palaeo.2007.12.002

Thiel M, Gutow L (2005a) The ecology of rafting in the marine environment: the floating substrata. Oceanogr Mar Biol 42:181–264

Thiel M, Gutow L (2005b) The ecology of rafting in the marine environment: the rafting organisms and community. Oceanogr Mar Biol 43:279–418. doi:http://dx.doi.org/10.1201%2F9781420037449.ch7

Thiel M, Haye PA (2006) The ecology of rafting in the marine environment: biogeographical and evolutionary consequences. Oceanogr Mar Biol 44:323–429. doi:http://dx.doi.org/10.1201%2F9781420006391.ch7

Valentine JW, Jablonski D (1985) Major determinants of the biogeographic pattern of the shallow-sea fauna. Bull Soc Géol France 24:893–899

Valentine J, Foin TC, Peart D (1978) A provincial model of Phanerozoic marine diversity. Paleobiology 4:55–66

Zazo C, Goy JL, Hillaire-Marcel C, Dabrio CJ, Gonzalez Delgado JA, Cabero A, Bardaji T, Ghaleb B, Soler V (2010) Sea level changes during the last and present interglacial in Sal Island (Cape Verde archipelago). Global Planet Change 72:302–317. doi:http://dx.doi.org/10.1016%2Fj.gloplacha.2010.01.006

# Chapter 7
# Morphometry

**Abstract** Morphological variations in bivalve shells are increasingly the focus of diverse studies that bridge palaeontology and ecology. Shape in bivalves is a key morphological characteristic that reflects both phylogenetic history and life habits. This chapter is centered on the different techniques that can be used to evaluate size and shape variability in Quaternary bivalves, and also gives reasons for such variations.

**Keywords** Southern South America · Quaternary · Pleistocene · Holocene · Mollusca · Morphometry · Linear morphometrics · Countor analysis · Landmarks

In relation to shell morphology, changes in space and time do not need to be interpreted solely as a species-level phenomenon, but can and should be considered in a community or palaeocommunity context, in which phenotypic variation between localities may represent a source of ecological information suitable for the evaluation of environmental changes.

Morphological variations in bivalve shells are increasingly the focus of diverse studies that bridge palaeontology and ecology. Shape in bivalves is a key morphological characteristic that reflects both phylogenetic history  and life habits (Stanley 1970; Crampton and Maxwell 2000).

## 7.1 Linear Morphometrics

Traditional linear morphometric analysis applied to bivalve remains is a potent tool for describing patterns of shell variation within species (Roy et al. 2001; Laudien et al. 2003). Gordillo et al. (2011b) used linear morphometric analysis to compare fossil and modern *Tawera gayi* shells. This species is a typical element of shallow marine soft bottoms in southern South America, and is the most common species recovered from Late Quaternary marine deposits along the Beagle Channel, in Tierra del Fuego. Two linear distances, shell length and shell height, were measured with a caliper in 304 unbroken *T. gayi* shells (194 modern

S. Gordillo et al., *Mollusk Shells as bio-geo-archives*,
South America and the Southern Hemisphere, DOI: 10.1007/978-3-319-03476-8_7,
© The Author(s) 2014

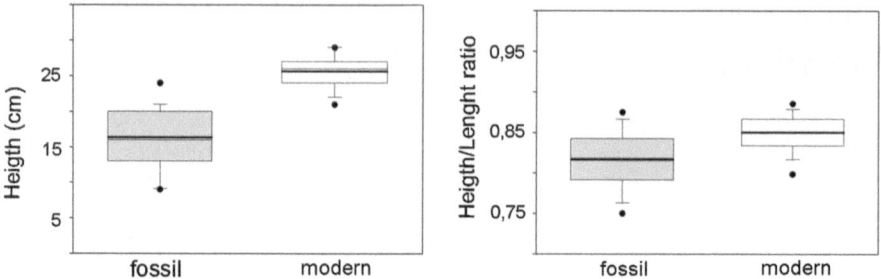

**Fig. 7.1** Boxplots showing the differences in size (*left*) and shape (*right*) between fossil and modern *Tawera gayi* shells. Modern shells reached larger sizes and are more quadrangular than fossil *T. gayi* shells (modified after Gordillo et al. 2011b)

and 110 fossil shells) from the Beagle Channel. Fossil shells previously dated at ca. 4,500–4,000 years were used for this analysis. The height/length ratio was used as a proxy for shape, and differences were evaluated with a nonparametric test. Results show that whereas the modern shells were more rounded, the fossil ones were slightly elongated. Fossil shells were also significantly smaller and shorter in length than modern shells (Mann–Whitney rank-sum test; $P < 0.001$ in both cases) (Fig. 7.1).

To discuss the possible causes of variation in shell morphology, these authors considered different ecological/environmental factors including temperature, productivity and biotic interactions.

Because *T. gayi* is a suspension feeder and is directly dependent on primary productivity for its growth, it is assumed here that increased primary productivity has a positive effect on shell growth. As modern *T. gayi* shells are larger than fossil shells, it is reasonable to infer that the increase in nutrient concentrations may have played a role in affecting shell size. Previous studies on bivalves (Kirby 2000; Vermeij 1990) and turritelid gastropods (Allmon 1992; Teusch et al. 2002) offer strong evidence that size and shape differences in shells may be explained by different temperature and productivity conditions. In southern South America, recent studies on venerids from Patagonia also show that shell variation is related to phenotypic plasticity as the result of different environmental conditions (Márquez et al. 2010).

However, higher productivity does not explain the different in shape between both ages, with modern shells more rounded than fossil ones. Other studies have indicated that morphological variations in mollusks may also result in biotic interactions such as predator–prey relationships (Hagadorn and Boyajian 1997; Teusch et al. 2002). Following on from this, Gordillo et al. (2011b) suggested that the reason modern shells were more rounded than fossil ones could be because it gave them a greater chance of avoiding drilling predation. This interpretation is supported by the fundamental relationship between shape and function in clams (Stanley 1975), and the development of antipredatory adaptation. Stanley (1975) observed that the prosogyrous condition and the rotational mechanism of burrowing are fundamental adaptations of burrowing clams, showing that each rocking motion of a typical clam involves a purely rotational movement, with no translational component. Thus, it could be predicted that the prosogyrous shape and flattened lunule should cause a backward rotation, shifting the axis of rotation towards the anterior region. The relationship between the length axis and the height axis (height/length ratio) therefore has a significant effect on the burrowing of clams, and the more rounded modern *T. gayi* shells offer less resistance to the substrate than the more elongated fossil *T. gayi* shells. Consequently, the more rounded clams burrow faster in order to avoid predation by muricid gastropods such as *T. geversianus* or *X. muriciformis*. To support this statement, a slight decrease in predation rate by drilling gastropods over time has been noted (Gordillo 1994, 1998). This is even more relevant if it is considered together with the short South American biogeographical history of *Tawera*, which apparently arrived from New Zealand during the Quaternary (see Sect. 6.3), and the need to improve strategies for avoiding predators in its new environment in South America. More work is needed to reinforce these assumptions that correlate changes in *T. gayi* shape with an effective resistance adaptation against drilling by gastropods (antipredatory adaptation).

## 7.2  Contour Analysis

Previous studies (Palmer et al. 2004; Rufino et al. 2006; Márquez et al. 2010, among others) have shown that Elliptic Fourier Analysis (EFA) on outline bivalve shells is very useful for defining specific shape features that might distinguish species or intraspecific variations from among different populations along a wide geographical range.

For the Quaternary of southern South America the contour method has been applied using fossil shells of three different infaunal clams; i.e., *Amiantis purpurata*, *Ameghinomya antiqua* and *Tawera gayi*.

Gordillo et al. (2011a), for example, analyzed the significance of the overall shell shape of *T. gayi* from different regions within the Magellan Region. Taking into account the palaeontological history of this genus in the southern hemisphere, EFA was also performed on shells of *Tawera* congeners from South Africa (*T. philomela*) and New Zealand (*T. spissa*). The use of EFA permitted the distinction between the three *Tawera* species and geographical differentiation in the *T. gayi* groups. The authors conclude that morphological variations of *T. gayi* appear best related to ecophenotypic plasticity as a response to different environmental conditions, although the palaeobiogeographical history of *Tawera* in South America cannot be ruled out.

Recently, Boretto et al. (unpublished data) analyzed changes in shape between Pleistocene, Holocene and Modern *Ameghinomya antiqua* shells from Bustamante Bay, in Patagonia (Argentina), using contour analysis (Fig. 7.2).

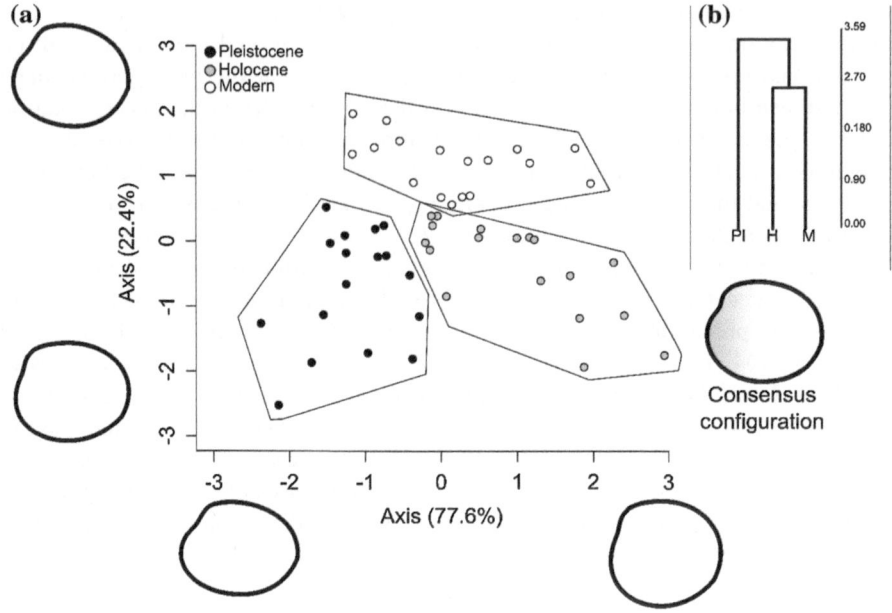

**Fig. 7.2** Contour method used on *Ameghinomya antiqua* from Quaternary marine deposits in Patagonia. **a** Analysis of shell shape variation along the first two canonical axes, plus diagrams of the reconstructed extreme configurations. **b** Cluster analysis showing the relationship of shell shapes between different ages

## 7.3  Landmarks

A third method to evaluate morphological changes is the analysis of landmarks. When applied to bivalves, this method offers a powerful technique for detecting differences between groups or for analyzing intra-specific variations between different populations (Rufino et al. 2006; Márquez et al. 2010).

Boretto et al. (unpublished data) also analyzed changes in shape between fossil and modern *Ameghinomya antiqua* shells from Bustamante Bay using 13 landmarks (Fig. 7.3), following Márquez et al. (2010).

This methodology, in addition to contour method, successfully separated Pleistocene ovoid shell shapes from rounded and sub-quadrangular shapes from the Holocene/Modern shells respectively. The authors explained these differences observed through time between Pleistocene and Holocene/Modern as the outcome of phenotypic plasticity under different environmental conditions, as previously mentioned by Márquez et al. (2010) for living populations.

During the Pleistocene, the configuration of Bustamante Bay might have been different from Holocene and Modern configurations, and the marine transgression/regression cycles might have modified the coastline evolution, including the development of a peninsula (Graviña) and the entire bay. Climatic oscillations

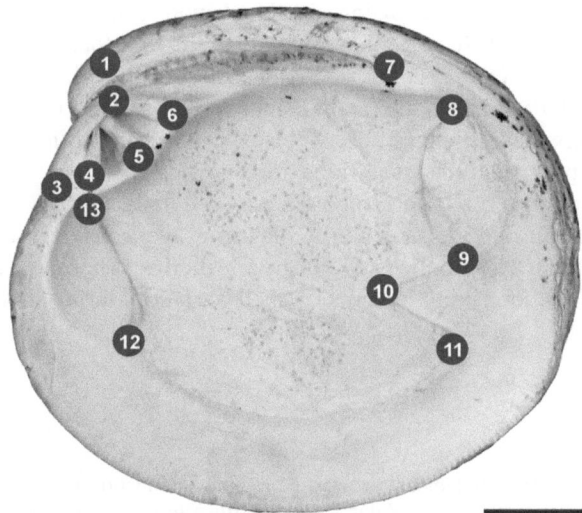

**Fig. 7.3** Diagram showing the *Ameghinomya antiqua* shell outline (*dashed line*) and the position of the 13 landmarks used to define the inner surface of the *right shell shape*. These landmarks are: *1*, the tip of the dorsal hinge ligament, *2* the tip of the cardinal teeth, *3* the lunule scar, *4* the end of the anterior cardinal tooth, *5* the midline cardinal tooth, *6* the end of the posterior cardinal tooth, *7* the tip of the posterior hinge ligament, *8–9* the anterior adductor muscle scar, *10* the tip of the pallial sinus, *11* the lower margin of the pallial sinus, *12–13* the posterior adductor muscle scar

could have affected the bottom sediments, and the chemical characteristics of water masses would have been influenced by continental melting. These factors, along with ocean circulation, were associated with environmental variation, and changes in the shell morphology of *A. antiqua* show how its phenotypic plasticity allowed it to adapt to different environments, including varied substrate, hydrological changes and predators. Thus, differences between Pleistocene and Holocene/Modern shells are believed to follow substratum changes along with ocean circulation. These changes most probably took place between the Last Glacial Maximum (24 kya BP; Rabassa 2008), when the sea level was 120 m below the present level, and the time of formation of the San Jorge Gulf (15 kya BP; Ponce et al. 2011).

On the other hand, Boretto et al. also consider other factors to explain differences between Holocene and Modern shells from Bustamante Bay in association with burial speed, which was described by Stanley (1970, 1975). The Holocene shells are more rounded and prosogyrous than the Modern specimens, and this condition may indicate a faster burial in the Holocene samples, which also burrow faster than the more elongated Pleistocene ones. A second feature to consider is the burrowing depth of bivalves, which can be regarded as the length of the siphon at its maximum extension in many cases (Kondo 1987, 1997; Zwarts and Wanink 1989). The pallial sinus, seen in the inner surface of siphonate bivalves, represents the accommodation space of a contracted siphon, and has long been regarded as an approximate measure of burrowing depth (Kondo 1987). In this respect, the Holocene/Modern shells have a longer mark than the Pleistocene shells (landmarks 9, 10, 11), and this feature supports the idea of deeper burial. When comparing Holocene and Modern shells, the pallial sinus length is more marked in the fossil samples. Perhaps these changes are associated with changes in the substratum or with predator–prey relationships which force clams to escape quickly. In connection with this interpretation ongoing research indicates a very low rate of drilling predation in the Pleistocene samples from the same region.

Thus, morphometrics applied to Quaternary mollusk shells is also a powerful analytical tool for describing patterns of shell variation during this period.

# References

Allmon WD (1992) Role of temperature and nutrients in extinction of turriteline gastropods: Cenozoic of the northwestern Atlantic and northeastern Pacific. Palaeogeogr Palaeocl 92:41–54. doi:http://dx.doi.org/10.1016/2F0031-0182/2892/2990134-Q

Crampton JS, Maxwell PA (2000) Size: all it's shaped up to be? evolution of shape through the lifespan of the Cenozoic bivalve *Spissatella* (Crassatellidae). In: Harper EM, Taylor JD, Crame JA (eds) Evolutionary biology of the Bivalvia. Geol Soc, London, Special Publ 177:399–423

Gordillo S (1994) Perforaciones en bivalvos subfósiles y actuales del Canal Beagle, Tierra del Fuego. Ameghiniana 31:177–0185

Gordillo S (1998) Trophonid gastropod predation on recent bivalves from the Magellanic Region. In: An eon of evolution. Paleobiological studies honoring Norman N. Newell, University of Calgary Press, Canada, pp 251–254

Gordillo S, Márquez F, Cárdenas J, Zubimendi MA (2011a) Shell variability in *Tawera gayi* from southern South America: a morphometric approach based on contour analysis. J Mar Biol Ass UK 91:815–822. doi:http://dx.doi.org/10.1017/2FS0025315410000391

Gordillo S, Martinelli J, Cárdenas J, Bayer S (2011b) Testing ecological and environmental changes during the last 6000 years: a multiproxy approach based on the bivalve *Tawera gayi* from southern South America. J Mar Biol Ass UK 91:1413–1427. doi:http://dx.doi.org/10.1017/2FS0025315410002183

Hagadorn JW, Boyajian GE (1997) Subtle changes in mature predator–prey systems: an example from Neogene Turritella Gastropoda. Palaios 12:372–379. doi:http://dx.doi.org/10.2307/2F3515336

Kirby MX (2000) Paleoecological differences between tertiary and quaternary *Crassostrea* oysters, as revealed by stable isotope sclerochronology. Palaios 15:132–141. doi:http://dx.doi.org/10.1669/2F0883-1351/282000/29015/3C0132/3APDBTAQ/3E2.0.CO/3B2

Kondo Y (1987) Burrowing depth of infaunal bivalves—observation f living species and its relation to shell morphology. Palaeont Soc Japan Trans Proc, NS 148:306–323

Kondo Y (1997) Inferred bivalve response to rapidal burial in a Pleistocene shallow-marine deposit from New Zealand. Palaeogeogr Palaeocl 128:87–100. doi:http://dx.doi.org/10.1016/2FS0031-0182/2896/2900039-9

Laudien J, Flint NS, Van Der Bank FH, Brey T (2003) Genetic and morphological variation in four populations of the surf clam *Donax serra* (Roding) from southern African sandy beaches. Biochem Syst Ecol 31:751–772. doi:http://dx.doi.org/10.1016/2FS0305-1978/2802/2900252-1

Márquez F, Robledo J, Escati Peñaloza G, Van der Molen S (2010) Use of different geometric morphometrics tools for the discrimination of phenotypic stocks of the striped clam

*Ameghinomya antiqua* (Veneridae) in San José Gulf, Patagonia, Argentina. Fish Res 101:127–131. doi:10.1016/j.fishres.2009.09.018

Palmer M, Pons G-X, Linde M (2004) Discriminating between geographical groups of a Mediterranean commercial clam (*Chamelea gallina* (L.): Veneridae) by shape analysis. Fish Res 67:93–98. doi:http://dx.doi.org/10.1016%2Fj.fishres.2003.07.006

Ponce JF, Rabassa J, Coronato A, Borromei AM (2011) Palaeogeographical evolution of the Atlantic coast of Pampa and Patagonia from the last glacial maximum to the Middle Holocene. Biol J Linn Soc 103:363–379. doi:http://dx.doi.org/10.1111/2Fj.1095-8312.2011.01653.x

Rabassa J (2008) Late Cenozoic glaciations in Patagonia and Tierra del Fuego. In: Rabassa, J (ed) The Late Cenozoic of Patagonia and Tierra del Fuego. Developments in Quaternary Science, vol 11. Elsevier, Amsterdam, p 151–204

Roy K, Jablonski D, Valentine J (2001) Climate change, species range limits and body size in marine bivalves. Ecol Lett 4:366–370. doi:10.1046/j.1461-0248.2001.00236.x

Rufino MM, Gaspar MB, Pereira AM, Vasconcelos P (2006) Use of shape to distinguish *Chamelea gallina* and *Chamelea striatula* (Bivalvia: Veneridae): linear and geometric morphometric methods. J Morphol 267:1433–1440. doi:10.1002/jmor.10489

Stanley SM (1970) Relation of shell form to life habits of the Bivalvia (Mollusca). Geol Soc Am, Inc. Mem 125:1–296. doi:http://dx.doi.org/10.1130/2FMEM125-p1

Stanley SM (1975) Why clams have the shape they have: an experimental analysis of burrowing bivalves. Paleobiology 1:48–58

Teusch K, Jones DS, Allmon WD (2002) Morphological variation in turritellid gastropods from the Pleistocene to Recent of Chile: association with upwelling intensity. Palaios 17:366–377. doi:http://dx.doi.org/10.1669/2F0883-1351/282002/29017/3C0366/3AMVITGF/3E2.0.CO/3B2

Vermeij GJ (1990) Tropical Pacific pelecypods and productivity: a hypothesis. B Mar Sci 47:62–67

Zwarts L, Wanink J (1989) Siphon size and burying depth in deposit and suspension-feeding benthic bivalves. Mar Biol 100:227–240. doi:http://dx.doi.org/10.1007/2FBF00391963

# Chapter 8
# Stable Isotopes

**Abstract** Carbonate shells have long been used as a proxy for paleoenvironmental conditions. In particular, oxygen and carbon stable isotopes from carbonate mollusk shells have been used for reconstructing water temperatures, the timing of upwelling events and changes in salinity. In this chapter, isotopic analysis was performed on shells of one particular species at different radiocarbon ages. It is also uses an example to show how to calculate paleotemperatures.

**Keywords** Southern South America · Quaternary · Holocene · Mollusca · Stable isotopes

Carbonate shells have long been used as a proxy for paleoenvironmental conditions. In particular, oxygen and carbon stable isotopes from carbonate mollusk shells have been used for reconstructing water temperatures, the timing of upwelling events and changes in salinity (e.g., Urey 1947; Epstein et al. 1951, 1953; Jones et al. 2005; Klein et al. 1996). $\delta^{18}O$ values of bivalve shell carbonate have been demonstrated to precipitate in equilibrium with the temperature and $\delta^{18}O$ value of the ambient water. Specifically, oxygen isotopic ratios can be converted to temperature values, and because oxygen is incorporated into calcium carbonate skeletons while the organism is growing, temperature curves can be constructed by sampling individuals from different sediments and ages. Water temperatures can therefore be reconstructed from $\delta^{18}O$ values of carbonate and an estimated $\delta^{18}O$ value of water, using an appropriate oxygen isotope carbonate fractionation relationship. By comparing oxygen and carbon isotopes from the same organism, productivity conditions may also be reconstructed (Jones and Allmon 1995). However, carbon isotopes can be more difficult to interpret than oxygen isotopes. The $\delta^{13}C$ value of marine shells is controlled by the $\delta^{13}C$ value of dissolved inorganic carbon found in the organism's internal water pool at the site of calcification (McConnaughey et al. 1997). Despite this, both oxygen and carbon have been shown to be useful tools for determining relative temperature and productivity conditions during different time periods.

S. Gordillo et al., *Mollusk Shells as bio-geo-archives*,
South America and the Southern Hemisphere, DOI: 10.1007/978-3-319-03476-8_8,
© The Author(s) 2014

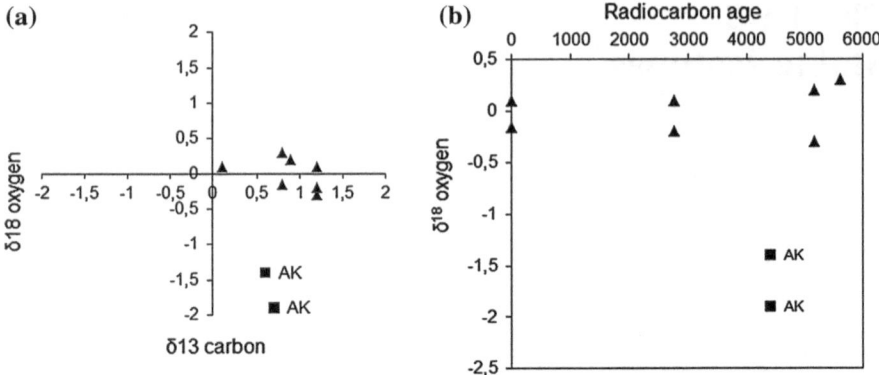

**Fig. 8.1** Stable isotopes of modern and fossil *Tawera gayi* shells from the Beagle Channel. **a** Relationship between oxygen and carbon isotopic composition in *T. gayi* shells, **b** Scatter plot of oxygen isotopic values of *T. gayi* shells, and their radiocarbon age. The great isotopic differences between the locality AK (*squares*) (Alakush; ca. 4400 years) and the other sites (*triangles*) is thought to be associated with warmer temperatures and a high volume of freshwater entering the Beagle Channel during the Hypsithermal (modified after Gordillo et al. 2011)

## 8.1  Hydrological Variations and Climatic Changes During the Holocene

Gordillo et al. (2011) explained changes in oxygen and carbon isotopes of *T. gayi* shells through hydrological changes and warmer temperatures during the Hypsithermal (Fig. 8.1).

Carbon values are within the −2 ‰ and +2 ‰ interval, which is associated with marine environments (Keith et al. 1964). In comparison, water coming from rivers is relatively deficient in $^{18}$O and $^{13}$C and is isotopically more variable: $\delta^{18}$O, −2 ‰; $\delta^{13}$C, 0 ‰ (Epstein and Mayeda 1953; Keith et al. 1964).

Similarly, the oxygen isotopic analysis of *T. gayi* shells (with the exception of shells from the Alakush site) showed values within a similar range. This isotopic data is difficult to explain (particularly carbon) because shells from the shallow marine environments along the Beagle Channel coast were exposed to freshwater, which derived from glacial ice melting and discharging into rivers that lead into the sea. In this respect, as shell carbonate is controlled by temperature and by the isotopic composition of ambient water, the stable isotopic composition of mollusk shells from freshwater environments shows wider and more depleted values than those from marine environments. This is due to the relative deficiency of $\delta^{18}$O and $\delta^{13}$C and the isotopically more variable nature of freshwater (see Wang et al. 1991). In the Magellan Region the mixture of seawater and freshwater from melting Andean snow also produces cooler waters (Palma and Aravena 2001). However, the great isotopic differences between Alakush (ca. 4400 years BP) and the other sites could be associated with warmer temperatures during the Hypsithermal (Obelic et al. 1998; Strelin et al. 2008) and a high volume of freshwater entering the Beagle Channel, partly due to an increase in rain (Candel et al. 2009), and partly to melting snow. During this period, large volumes of water enriched in $^{16}O$ flowed into the sea, thus resulting in lower ratios.

The carbon isotopic analysis of *T. gayi* shells indicates the existence of a mixing of waters from pure marine waters to marine waters with signs of fresh-water influence. The high depletion of $\delta^{18}O$ at ca. 4400 years BP would be associated with warmer temperatures during the Hypsithermal, and a maximum freshwater input to the Beagle Channel, probably due to an increase in rain during this period.

## 8.2 Age Calibration and Temperature Equations

Carbonate skeletal remains of mollusks can provide reliable data on environmental parameters, and combined with $^{14}C$ dating information they can also give us valuable data on paleotemperatures.

We therefore used the oxygen composition of radiocarbon-dated *Mytilus* shells from the Beagle Channel to infer paleotemperatures during the Holocene. *Mytilus* shells were chosen for oxygen isotope analysis because it is the only species available throughout the Holocene, and appears in the oldest, as well as the youngest, marine records. One disadvantage of this species is that most of the specimens recovered were broken (due to taphonomic reasons), which prevented sclerochronological analysis on complete specimens. The entire shell is therefore used to estimate an 'average' temperature at the age of deposition. The principle applied here is that bivalve mollusks living at the same time and in the same place will experience the same temperature and will therefore have similar oxygen isotope mean values.

To perform isotopic analysis, a total of 62 *Mytilus* shells were obtained from nine paleontological sites previously dated. In the laboratory at the Earth Science Department, University of Pisa (Italy), shells were crushed to a fine powder and analyzed using standard methods.

Conventional $^{14}C$-ages were calibrated using the CalPal program. As a first approximation towards correcting these dates for the marine reservoir effect, a marine calibration dataset (Reimer and Reimer 2000) was used which incorporates a time-dependent global ocean reservoir correction of about 400 years, and a difference $\Delta R$ of $221 \pm 40$ in reservoir age (to accommodate the local effects of this region), in conjunction with the CalPal program for calibrating these samples (Fig. 8.2).

To estimate an average temperature at the age of deposition, water temperatures were calculated using a paleotemperature equation (Eq. 8.1) introduced by Epstein and Mayeda (1953) and improved by Craig (1965).

$$T(^{\circ}C) = 16.9 - 4.2 * (\delta_c - \delta_w) + 0.13 * (\delta_c - \delta_w)^2 \qquad (8.1)$$

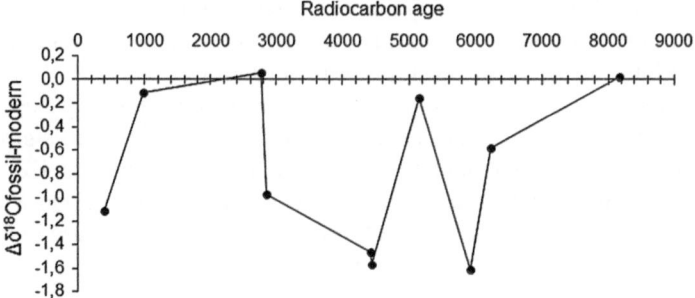

**Fig. 8.2** Variations between oxygen isotope values in *Mytilus* fossil specimens of different radiocarbon ages. Each dot in the graph represents the difference between the oxygen isotope value of a living and a fossil specimen of a different [14]C age. The line shows the trend through time

In this equation $T$ represents temperature (°C). Both $\delta_c$ and $\delta_w$ are expressed as $^{18}O/^{16}O$ isotope ratios. $\delta_c$ represents the oxygen isotopic composition of the carbonate expressed as a deviation in parts per thousand from a standard carbonate (i.e., PeeDee Belemnite, a carbonate fossil from South Carolina). $\delta_w$ represents the oxygen isotopic composition of the water expressed as deviation from standard mean ocean water (SMOW). Based on only one available record for the Beagle Channel provided by Obelic et al. (1998), a value of $\delta^{18}O_w = -1.81$ ‰ PDB was used, measured at Ushuaia Bay with a salinity of around 30 PSU.

In this equation a theorical average value of $\delta^{18}O_w = -1.09$ ‰ PDB for 31 PSU was used, calculated on the basis of the relationship between two pieces of isotopic data: on the one hand, a mean value of $\delta^{18}O_w = -10.75$ ‰ PDB, obtained from precipitation by Iturraspe et al. (1989), equivalent to a freshwater input with 0 PSU; and on the other hand, a value of $\delta^{18}O_w = -0.16$ ‰ PDB, obtained from surface salinity of 34 PSU by Meredith et al. (1999) in the Drake Passage. A quite similar equation was used in Colonese et al. (2011).

A second equation (Eq. 8.2) is based on Wanamaker et al. (2007).

$$T(°C) = 16.28 - 4.57 * (\delta_c - \delta_w) + 0.06 * (\delta_c - \delta_w)^2 \tag{8.2}$$

In this study, it was assessed whether the $\delta^{18}O$ values of *Mytilus* shells from the Beagle Channel can be used as a proxy for paleotemperatures during the Holocene. Due to modern samples did not live at the same time and in the same place (i.e., they were both temporally and spatially mixed; see Goodwin et al. 2004), they did not experience the same temperature, and therefore had different oxygen values. However, fossil values represent an average of the same time/same place. Despite this, clear differences in mean temperatures were demonstrated within the period between 9,000 and 2,000 years, starting with colder conditions (2 °C below) ca. 8,500 years, then a warm peak at ca. 6,500 years (3 °C higher), followed by a second period of cold conditions at ca. 5,000 years BP and finally a second warmer period ca. 4,500–4,000 (2 °C higher).

Previous studies on bivalve shell carbon isotope compositions show that they are difficult to interpret, and are complicated by a number of factors including the contribution of metabolic carbon (Gillikin et al. 2006). However, the paleotemperature curve agrees well with previous environmental interpretations made by other authors (e.g., Heusser 1998; Grill et al. 2002) who worked on other proxies such as palynology.

In Chap. 9 more isotopic data, combined with individual growth using sclerochronology, and calibrated against temperatures, is used to reinforce evidence of the impact of climatic changes on shell growth and structure, and to discriminate better between environmental changes and ecological reasons.

# References

Candel MS, Borromei MS, Martínez MA, Gordillo S, Quattrocchio M, Rabassa JO (2009) Middle-Late Holocene palynology and marine mollusks from Archipiélago Cormoranes area, Beagle Channel, southern Tierra del Fuego, Argentina. Palaeogeogr Palaeocl 273:111–122. doi:http://dx.doi.org/10.1016/2Fj.palaeo.2008.12.009

Craig H (1965) The measurement of oxygen isotope palaeotemperatures. In: Tongiorgi E (ed) Stable isotopes in oceanographic studies and palaeotemperatures. Consiglio Nazionale della Richerche Laboratorio di Geologia Nucleare, Pisa, pp 161–182

Colonese AC, Camarós E, Verdún E, Estévez J, Giralt S, Rejas M (2011) Integrated archaeozoological research of shell middens: new insights into hunter-gatherer-fisher coastal explotation in Tierra del Fuego. J Island Coast Arch 6:235–254. doi:http://dx.doi.org/10.1080%2F15564894.2011.586088

Epstein S, Mayeda T (1953) Variation of O18 content of waters from natural sources. Geochim Cosmochim Ac 4:213–224. doi:http://dx.doi.org/10.1016%2F0016-7037%2853%2990051-9

Epstein S, Buchsbaum R, Lowenstam HA, Urey HC (1951) Carbonate-water isotopic temperature scale. Bull Geol Soc Am 62:417–426. doi:http://dx.doi.org/10.1130/2F0016-7606/281951/2962/5B417/3ACITS/5D2.0.CO/3B2

Epstein S, Buchsbaum R, Lowenstam HA, Urey HC (1953) Revised carbonate-water isotopic temperature scale. Bull Geol Soc Am 64:1315–1326. doi:http://dx.doi.org/10.1130/2F0016-7606/281953/2964/5B1315/3ARCITS/5D2.0.CO/3B2

Gillikin DP, Lorrain A, Bouillon S, Willenz P, Dehairs F (2006) Stable carbon isotopic composition of *Mytilus edulis* shells: relation to metabolism, salinity, d13CDIC and phytoplankton. Org Geochem 37:1371–1382. doi:10.1016/j.orggeochem.2006.03.008

Goodwin DH, Flessa KW, Tellez-Duarte MA, Dettman DL, Schöne BR, Avila-Serrano GA (2004) Detecting time-averaging and spatial mixing using oxygen isotope variation: a case study. Palaeogeogr Palaeocl 205:1– 21. doi:http://dx.doi.org/10.1016%2Fj.palaeo.2003.10.020

Gordillo S, Martinelli J, Cárdenas J, Bayer S (2011) Testing ecological and environmental changes during the last 6000 years: a multiproxy approach based on the bivalve *Tawera gayi* from southern South America. J Mar Biol Ass UK 91:1413–1427. doi:http://dx.doi.org/10.1017/2FS0025315410002183

Grill S, Borromei AM, Quattrocchio M, Coronato A, Bujalesky G, Rabassa J (2002) Palynological and sedimentological analysis of recent sediments from Río Varela, Beagle Channel, Tierra del Fuego, Argentina. Rev Esp Micropaleont 34:145–161

Heusser CJ (1998) Deglacial paleoclimate of the American sector of the Southern Ocean: late Glacial-Holocene records from the latitude of Beagle Channel (55°S), Argentine Tierra del Fuego. Palaeogeogr Palaeocl 141:277–301

Iturraspe R, Sottini R, Schroder C, Escobar J (1989) Hidrología y variables climáticas del Territorio de Tierra del Fuego. Contribución Científica del Centro Austral de Investigaciones Científicas 7:1–196

Jones DS, Allmon WD (1995) Records of upwelling, seasonality and growth in stable isotope profiles of Pliocene mollusk shells from Florida. Lethaia 28:61–74. doi:http://dx.doi.org/10.1111%2Fj.1502-3931.1995.tb01593.x

Jones DS, Quitmyer IR, C Fred T Andrus (2005) Oxygen isotopic evidence for greater seasonality in Holocene shells of Donax variabilis from Florida. Palaeogeogr Palaeocl 228:96–108. doi:http://dx.doi.org/10.1016/2Fj.palaeo.2005.03.046

Keith ML, Anderson GM, Eichler R (1964) Carbon and oxygen isotopic composition of mollusk shells from marine and fresh-water environments. Geochim Cosmochim Ac 28:1757–1786. doi:http://dx.doi.org/10.1016/2F0016-7037/2864%2990021-3

Klein RT, Lohmann KC, Thayer CW (1996) Sr/Ca and 13C/12C ratios in skeletal calcite of *Mytilus trossulus*: covariation with metabolic rate, salinity, and carbon isotopic composition of seawater. Geochim Cosmochim Ac 60:4207–4221

Meredith MP, Grose KE, McDonagh EL, Heywood KJ, Frew RD, Dennis P (1999) The distribution of oxygen isotopes in the water masses of Drake Passage and the South Atlantic. J Geophys Res 104:20949–20962. doi:10.1029/98JC02544

McConnaughey T, Burdett J, Whelan JF, Paull CK (1997) Carbon isotopes in biological carbonates: respiration and photosynthesis. Geochim Cosmochim Ac 61:611–622. doi:http://dx.doi.org/10.1016%2FS0016-7037%2896%2900361-4

Obelic B, Álvarez A, Argullós J, Piana EL (1998) Determination of water palaeotemperature in the Beagle Channel (Argentina) during the last 6000 yr through stable isotope composition of *Mytilus edulis* shells. Quat S Am Ant Pen 11:47–71

Palma S, Aravena G (2001) Distribución de quetognatos, eufáusidos y sifonóforos en la Región Magallánica. Cienc Tecn Mar 24:47–49

Strelin J, Casassa G, Rosqvist G, Holmlund P (2008) Holocene glaciations in the Ema Glacier valley, Monte Sarmiento Massif, Tierra del Fuego. Palaeogeogr Palaeocl 260:299–314. doi:http://dx.doi.org/10.1016/2Fj.palaeo.2007.12.002

Urey HC (1947) The thermodynamic properties of isotopic substances. J Chem Soc: 562–581. doi:http://dx.doi.org/10.1039/2Fjr9470000562

Wanamaker AD Jr, Kreutz KJ, Borns HW Jr, Introne DS, Feindel S, Funder S, Rawson PD, Barber BJ (2007) Experimental determination of salinity, temperature, growth, and metabolic effects on shell isotope chemistry of *Mytilus edulis* collected from Maine and Greenland. Paleoceanography 22: PA2217. doi:10.1029/2006PA001352

Reimer R, Reimer R (2000) Marine reservoir correction database http://calib.org/marine

Wang CH, Peng TR, Chen PF (1991) Oxygen and carbon isotopic compositions of mollusks from the Late Pleistocene Szekou Formation, Southern Taiwan. P Natl Sci Coun, Rep Chin 5:455–464

# Chapter 9
# Sclerochronology

**Abstract** As mollusks grow, their shells become biogeochemical records of the environmental and climatic conditions experienced throughout their lifetime. Following the ideas developed in Chap. 8, in this chapter stable isotopes are treated in conjunction with individual growth in the selected fossil specimens of different bivalves.

**Keywords** Southern South America · Quaternary · Pleistocene · Holocene · Mollusca · Stable isotopes · Individual growth · Annual growth · Sclerochronology

As mollusks grow, their shells become biogeochemical records of the environmental and climatic conditions experienced throughout their lifetime (for a review see Rhoads and Lutz 1980). The kind of study that focuses primarily upon physicochemical variations in the periodically hard tissues of organisms has been defined as sclerochronology (Oschmann 2009). Research over the last decade has demonstrated that bivalves are a very valuable tool for reconstructing environmental conditions in both extant and ancient marine environments (e.g., Schöne et al. 2004, 2005; Watanabe et al. 2004; Carré et al. 2005, among others). Additionally, their broad biogeographic distribution enables cross-calibration with other environmental and climate archives (Schöne and Gillikin 2013).

To obtain seasonal records of change in a region, it is possible to recover proxy climate data from the carbonate of marine bivalve shells obtained from the different environments within the studied area (Fig. 9.1).

Temperature seasonality, the difference between summer and winter temperatures, is one of the most important characteristics of climate, and plays a crucial role in determining the surface characteristics of the ocean. In our study we used $\delta^{18}O$ values of microdrilled bivalves (Fig. 9.2) to reconstruct series (Fig. 9.3) in order to evaluate seasonal temperature variations at discrete intervals of 2 and more years each from shells of different ages.

S. Gordillo et al., *Mollusk Shells as bio-geo-archives*,
South America and the Southern Hemisphere, DOI: 10.1007/978-3-319-03476-8_9,
© The Author(s) 2014

**(a)**                                                     **(b)**

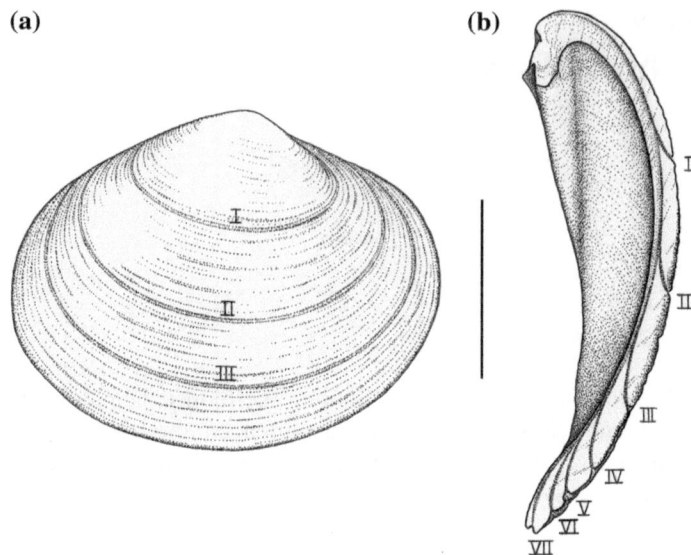

**Fig. 9.1** Annual growth lines in a typical bivalve shell. **a** Annual growth lines (*I, II, III*) observed through an external view. **b** A *cross section* of **a** showing the same lines seen in **a**, and also other annual growth lines (*IV, V, VI, VII*) located near the marginal zone and only visible through a cross section

In southern South America little is known about the biotic response of marine individual species to large scale climate variability throughout the Holocene. One exception is a recent study by Colonese et al. (2012), centered on the analysis of stable oxygen isotope values from shells of the limpet *Nacella deaurata*, which were recovered from archaeological shell middens located along the Beagle Channel. This study suggests that animals were collected in winter and that they experienced similar environmental conditions to the present day conditions at ca. 1320 years BP.

Fossil shells of the aragonitic bivalve *Retrotapes exalbidus* provide us with the opportunity to investigate climate variability in the Beagle Channel, as well as past seasonal dynamics of sea water temperature during the mid-to-late-Holocene. This selection is based on two reasons: extant *R. exalbidus* preserves annual increments in the outer shell layer, thus capturing the full seasonal temperature amplitude in its shell (Lomovasky et al. 2002; Yan et al. 2012); and, although not as common as other venerids, this species is well preserved in different Holocene marine outcrops along the channel (Gordillo et al. 2005; Cárdenas and Gordillo 2009). For this purpose, Gordillo et al. (2013) performed a preliminary analysis of Holocene fossil *R. exalbidus* shells which were sectioned, polished, photographed and measured, and after examination three of them were selected for chemical sampling. In each case, one-half of the shell was used to resolve the annual growth bands and the other

**Fig. 9.2** Sampling methods of carbonate powder for isotopic analysis. **a** Sampling on the *outer* surface of a shell (external shell sampling), **b** Sampling from a *cross sectioned* shell (internal shell sampling)

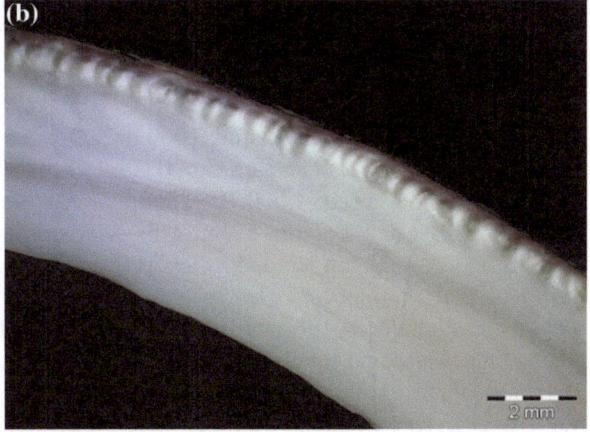

half was used for stable isotope sampling. In addition, a fragment of each shell was used for $^{14}C$ dating performed by the accelerator mass spectrometry (AMS) technique in the Poznań Radiocarbon Laboratory. Ontogenetic ages were measured by counting the annual growth increments under a stereo microscope. The results show differences between the three specimens. In the ontogenetic oldest individual (14 years old), which had a calibrated mean value age of 3839 BP, the $\delta^{18}O$ values ranged from 1.53 to −1.16 ‰. Another younger specimen (8 years old), with a mean calibrated age of 431 BP, had $\delta^{18}O$ values from 1.55 to 0.44 ‰. A third specimen (10 years old), with a calibrated age of 5190 BP, had $\delta^{18}O$ values from 1.29 to 0.72 ‰. Variations in annual growth increment widths were also found at different radiocarbon ages, probably correlated with environmental changes over the mid-to-late Holocene. The most positive $\delta^{18}O$ values were correlated with winter and the most negative $\delta^{18}O$ with summer. In addition, the summer values around 3800 years

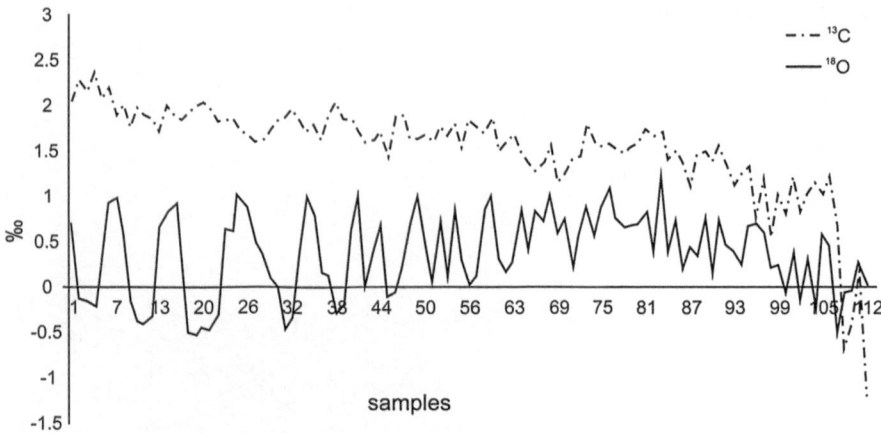

**Fig. 9.3** Oxygen and carbon chronologies from a (middle-late) Holocene shell of *Amiantis purpurata*. Each sample (x axis) presented $\delta^{18}O$ and $\delta^{13}C$ (y axis) values expressed by parts per thousand (‰) (Bayer et al. 2013)

BP are more negative than around 5,000 or 500 years BP. These findings correlate well with a cooling episode at ca. 5,000 BP, followed by a period of ameliorization (the Hypsithermal at ca. 4,000 BP), and towards the end of the Holocene, at ca. 500 years BP, a new cooling event. This sclerochronological study of the growth patterns and the oxygen isotope ratios in fossil *R. exalbidus* shells therefore demonstrated that this species clearly exhibited annual cycles showing seasonality patterns through the mid-to-late Holocene, thus providing us with an opportunity to analyze intra-seasonal time scales in the fossil record.

In another preliminary report, Lomovasky et al. (2013) analyzed fossil and modern *Tawera gayi* shells from the Beagle Channel. The $\delta^{18}O$ values obtained in fossil shells ranged from 1.316 to –0.064 ‰. They correlated the most positive $\delta^{18}O$ values with winter forming translucent bands and the most negative $\delta^{18}O$ with summer. The authors demonstrated that this species clearly exhibited annual cycles which showed seasonality patterns from the mid-Holocene to the present, with translucent bands corresponding to slow or halted growth formed in fall/winter. The growth rate of *Tawera gayi* was also lower during the past warm period (Hypsithermal) than the present, which is possibly related to different productivity in the channel and/or a lower metabolic rate of the clams when exposed to a higher temperature.

Given these preliminary results, it can be seen that sclerochronology is a promising path to explore further in (Fig. 9.4).

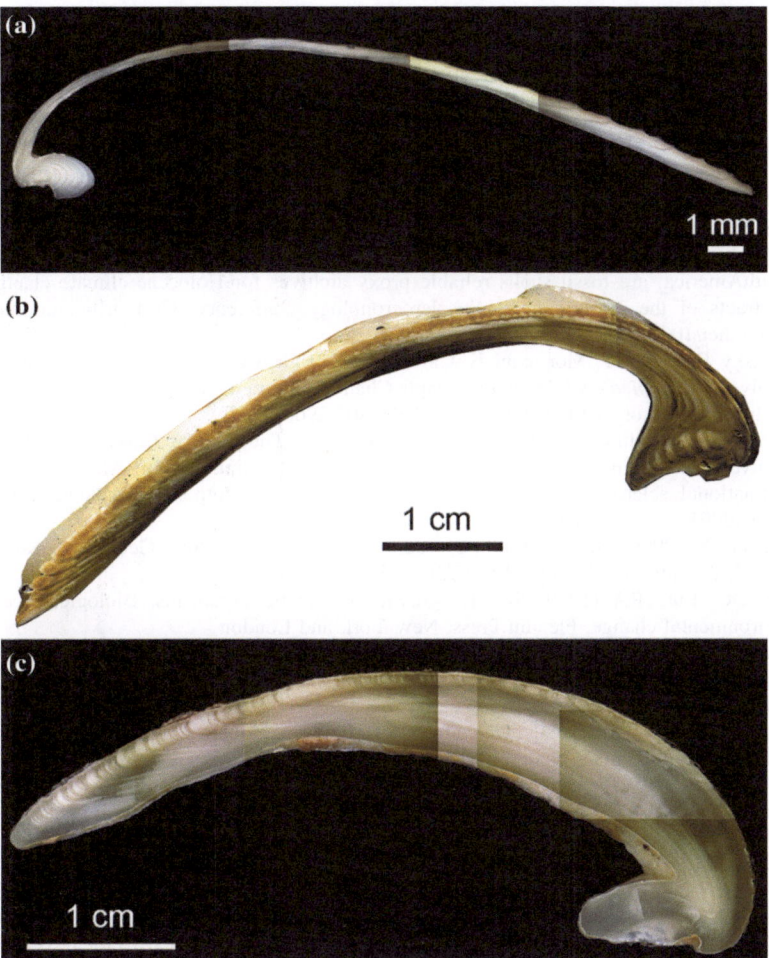

**Fig. 9.4** *Thin* sections of shells of different venerid taxa showing annual increments. **a** *Retrotapes exalbidus* (Holocene), **b** *Ameghinomya antiqua* (Pleistocene), **c** *Amiantis purpurata* (Pleistocene)

# References

Bayer MS, Brey T, Beierlein L, Gordillo S (2013) Late quaternary climatic variability in northern Patagonia Argentina—information from modern and fossil shells of *Amiantis purpurata* (Bivalvia, Veneridae). In: Abstracts of the 3rd international sclerocronology conference, Caernarfon. doi:http://hdl.handle.net/10013/epic.41694

Cárdenas J, Gordillo S (2009) Paleoenvironmental interpretation of late Quaternary molluscan assemblages from southern South America: a taphonomic comparison between the Strait of Magellan and the Beagle Channel. Andean Geol 36:81–93. doi:http://dx.doi.org/10.4067/2FS0718-71062009000100007

Carré M, Bentaleb I, Blamart D, Ogle N, Cardenas F, Zevallos S, Kalin RM, Ortlieb L, Fontugne M (2005) Stable isotopes and sclerochronology of the bivalve *Mesodesma donacium*:

potential application to Peruvian paleoceanographic reconstructions. Palaeogeogr Palaeocl 228:4–25. doi:http://dx.doi.org/10.1016%2Fj.palaeo.2005.03.045

Colonese AC, Verdún-Castelló E, Álvarez M, Godino IB, Zurro D, Salvatelli L (2012) Oxygen isotopic composition of limpet shells from the Beagle Channel: implications for seasonal studies in shell middens of Tierra del Fuego. J Archaeol Sci 39:1738–1748. doi:http://dx.doi.org/10.1016%2Fj.jas.2012.01.012

Gordillo S, Coronato A, Rabassa J (2005) Quaternary molluscan faunas from the island of Tierra del Fuego after the Last Glacial Maximum. Sci Mar 69 (Suppl. 2):337–348. doi:10.3989/scimar.2005.69s2337

Gordillo S, Brey T. Beyer K, Lomovasky B (2013) *Retrotapes exalbidus* from southern SouthAmerica: are fossil shells reliable proxy archives for Holocene climate changes? In: Abstracts of the 3rd international sclerocronology conference, Caernarfon. doi:http://hdl.handle.net/10013/epic.41695.d001

Lomovasky BJ, Brey T, Morriconi E, Calvo J (2002) Growth and production of the venerid bivalve *Eurhomalea exalbida* in the Beagle Channel, Tierra del Fuego. J Sea Res 48:209–216. doi:http://dx.doi.org/10.1016/2FS1385-1101/2802/2900133-8

Lomovasky, B, Gordillo, S, Alvarez, G, Brey, T (2013) The bivalve *Tawera gayi*, a potential archive of southern South America Holocene climate variability. In: Abstracts of the 3rd international sclerocronology conference, Caernarfon. doi:http://hdl.handle.net/10013/epic.41693.d001

Oschmann W (2009) Sclerochronology: editorial. Int J Earth Sci (GeologischeRundschau) 98:1–2. doi:http://dx.doi.org/10.1007/2Fs00531-008-0403-3

Rhoads DC, Lutz RA (1980) Skeletal growth of aquatic organisms. Biological records of environmental change. Plenum Press, New York and London

Schöne BR, Gillikin DP (2013) Unraveling environmental histories from skeletal diaries: advances in sclerochronology. Palaeogeogr Palaeocl 373:1–5. doi:http://dx.doi.org/10.1016%2Fj.palaeo.2012.11.026

Schöne BR, Oshmann W, Tanabe K, Dettman DL, Fiebig J, Houk SD, Kanie Y (2004) Holocene seasonal environmental trends at Tokyo Bay, Japan, reconstructed from bivalve mollusk shells- implication for changes in the East Asian monsoon and latitudinal shifts of the Polar Front. Quaternary Sci Rev 23:1137–1150. doi:http://dx.doi.org/10.1016/2Fj.quascirev.2003.10.013

Schöne BR, Fiebig J, Pfeiffer M, Gleb R, Hickson J, Johnson A, Dreyer W, Oschmann W (2005) Climate records from a bivalve Methuselah (*Arctica islandica*, Mollusca; Iceland). Palaeogeogr Palaeocl 228:130–148. doi:http://dx.doi.org/10.1016%2Fj.palaeo.2005.03.049

Watanabe T, Suzuki A, Kawahata H, Kan H, Ogawa S (2004) A 60-year isotopic record from a mid-Holocene fossil giant clam (Tridacna gigas) in the Ryukyu Islands: physiological and paleoclimatic implications. Palaeogeogr Palaeocl 212:343–354. doi:http://dx.doi.org/10.1016/2FS0031-0182/2804/2900358-X

Yan L, Schöne BR, Arkhipkin A (2012) *Eurhomalea exalbida* (Bivalvia): a reliable recorder of climate in southern South America? Palaeogeogr Palaeocl 350–352: 91–100. doi:http://dx.doi.org/10.1016/2Fj.palaeo.2012.06.018

# Chapter 10
# Concluding Remarks

**Abstract** In this chapter we summarize all the information presented so far, and explain why we consider mollusk shells to be excellent geo-bio-archives for the reconstruction of Quaternary paleoenvironments and for paleoclimatic interpretations.

**Keywords** Southern South America · Quaternary · Pleistocene · Holocene · Mollusca · Multidisciplinary approach · Faunistic changes · Local/regional scale

The Argentinean coastline makes an interesting case study due to its huge latitudinal extension and the fact that the coastline is shaped by marine-terrace deposits accumulated throughout the Quaternary; this provides us with an opportunity to collect and examine molluscan death assemblages from modern environments and fossil marine deposits. In this respect, comparisons between the modern and adjacent fossil shells are probably the most effective way of evaluating changes and of reconstructing past environmental conditions during the most recent geological period.

Given the wide range of disciplines involved in Quaternary research, this book has aimed to provide a comprehensive multidisciplinary approach to how mollusk shell remains (mollusk assemblages and/or selected taxa) have been used in the reconstruction of Quaternary environments in southern South America. Our study was based on Present-day, Holocene and Pleistocene mollusk assemblages from different areas covering a wide distribution range between 40° and 54° S.

This book has emphasized the use of different proxies (taphonomy, paleoecology, morphometry, stable isotopes, sclerochronology) as a strategy for a better understanding of environmental and climatic changes, and how these changes are reflected in shells. For this purpose, different lines of evidence were applied at local/regional level in different locations along the Argentinean coast. This scale of analysis helped us to identify local/regional changes, in addition to other global events, for example, the Hypsithermal during the mid-Holocene and the Marine Isotope Stage 5e during the late Pleistocene, both slightly warmer periods than the present. However, more work is needed on local and regional levels to fully

S. Gordillo et al., *Mollusk Shells as bio-geo-archives*,
South America and the Southern Hemisphere, DOI: 10.1007/978-3-319-03476-8_10,
© The Author(s) 2014

understand which environmental changes are due to global causes and which are due to regional/local causes.

Although the Quaternary was a short geological period, notable events occurred. It is widely known that significant glacioeustatic movements took place during the different glacial/interglacial cycles of the Quaternary, with lower sea levels during colder periods. In southern South America, a large portion of the Argentinean continental shelf was exposed during these glacial periods, and was then quickly covered during the interglacial periods. The coastal environments would therefore have suffered a rapid migration of taxa as a response to the fast transgressive sea.

Signs of these changes have been preserved in local mollusk assemblages, which have also changed over time. At least four mollusk species became extinct during the Pleistocene, and at times during both the Pleistocene and the Holocene, several taxa shifted their range of distribution or colonized new vacant areas.

A local-scale quantitative and qualitative analysis of mollusk assemblages at different latitudes suggests that each environment changed over time, acting as a 'dynamic mosaic' for the development of local communities in patchy habitats or sub-environments which, in accordance with sea-level changes, also shifted over time. Changes in mollusk assemblages took place from a few hundred to several thousand years ago, and mostly follow local physical variations (i.e., substrate, availability of food and currents); changes associated to large scale climatic variability during the Holocene were also recorded in individual taxa.

The vision that emerges from the Quaternary mollusk shells in southern South America is a history of continuous and dynamic shifts of the local benthic communities in response to disturbances in physical conditions. Faunistic changes during the Quaternary partly reflect changes over time, and partly reflect local circumstances. Discerning how much of each is still difficult. One significant outcome of this work is that those analyses based solely on lists of species on a biogeographic scale should be treated with caution if used to assess changes in the Quaternary: local analyses are far more reliable and are more powerful tools.

Because the multi-proxy evidence used in this study provides a consistent picture of spatial and temporal environmental and climatic changes in southern South America, we believe that mollusk shells are extremely valuable tools for studies addressing Quaternary environments anywhere.

# Index

S. Gordillo et al., *Mollusk Shells as bio-geo-archives*,
South America and the Southern Hemisphere, DOI: 10.1007/978-3-319-03476-8,
© The Author(s) 2014